Springer Series in Pharmaceutical Statistics

Editors
F. Bretz
P. Müller
T. Permutt
J. Pinheiro

More information about this series at http://www.springer.com/series/15122

Geraldine Rauch • Svenja Schüler •
Meinhard Kieser

Planning and Analyzing Clinical Trials with Composite Endpoints

Springer

Geraldine Rauch
Institute of Biometry and Clinical Epidemiology
Charité - Universitätsmedizin Berlin
Berlin, Germany

Svenja Schüler
Institute of Medical Biometry and Informatics
University of Heidelberg
Heidelberg, Germany

Meinhard Kieser
Institute of Medical Biometry and Informatics
University of Heidelberg
Heidelberg, Germany

ISSN 2366-8695   ISSN 2366-8709 (electronic)
Springer Series in Pharmaceutical Statistics
ISBN 978-3-030-08855-2   ISBN 978-3-319-73770-6 (eBook)
https://doi.org/10.1007/978-3-319-73770-6

Mathematics Subject Classification (2010): 62L05, 62P10

Printed on acid-free paper

This Springer imprint is published by the registered company Springer International Publishing AG part of Springer Nature.
The registered company address is: Gewerbestrasse 11, 6330 Cham, Switzerland

We would like to thank our colleagues from the Institute of Medical Biometry and Informatics, University of Heidelberg, for many discussions on the topic of composite endpoints and for reviewing parts of the text, especially Eva Dölger, Ann-Kathrin Ozga, and Stella Preussler. In addition, we are grateful to Springer Publishers for the decision to publish this book. A special thanks goes to our editor Dr. Eva Hiripi for her continuous encouragement and support for this project.

Preface

Composite endpoints are often used as primary efficacy variables for clinical trials, particularly in the field of oncology and cardiology. These endpoints combine several variables of interest within a single composite measure. By this, all variables which are of major clinical relevance can be considered in the primary analysis without the need to adjust for multiplicity. Moreover, it is intended to enlarge the number of expected events and thereby to increase the power of the clinical trial. For the latter reason, composite endpoints are often employed if the variables of interest correspond to rather rare events. This concept can be illustrated by means of a fairytale. Each of the following animals—donkey, dog, cat, and cock—taken on its own has a relatively small height. Putting these animals on top of each other, they form the "Bremer Town Musicians" (Bremer Stadtmusikanten) which are now very large and impressive.

Coming back to real clinical trial applications in oncology and cardiology, the most relevant endpoint often corresponds to "death". However, if the survival prognosis of the patient population of interest is not too bad, then it might not be feasible to wait until an effect in the death rates can be observed. To resolve this problem, the outcome "death" might be combined with other disease-related events which occur more frequently. There exist some major challenges when using such a composite endpoint as primary efficacy variable. On the one hand, a serious difficulty in the planning stage is that the sample size calculation is based on more parameter assumptions as compared to a clinical trial with a single-variable primary endpoint. Thus, the target sample size is often subject to a high level of uncertainty. This is due to the fact that the assumed effect for the composite endpoint which is used for sample size calculation depends both on the effects in the single components and on the correlation between them. On the other hand, the interpretation of composite endpoints can be difficult, as the observed effect for the composite does not necessarily reflects the effects for the single components. Therefore, it might not be adequate to judge the efficacy of a new intervention exclusively based on the composite effect.

This book is structured into six parts. In Part I, the general concepts of composite endpoints are introduced. In Chap. 1, we begin by defining composite endpoints and by providing the rationale for the application of composite primary endpoints in clinical practice. In Chap. 2, the challenges resulting from the use of composite endpoints are introduced and discussed. Chapter 3 presents recommendations and open issues related to composite endpoints which are discussed by current guidelines in the field of clinical trial methodology, benefit-risk and health technology assessment, as well as by disease-specific guidelines. Part I concludes with Chap. 4 where an overview of some exemplary clinical trials is given which illustrate different aspects related to composite endpoints.

In Part II, we formulate the mathematical background of the underlying test problem. In this part of the book, we focus on a confirmatory test problem which is formulated for a single (composite) endpoint. The underlying test hypotheses, the test statistics, as well as strategies for sample size calculation are provided for composite binary endpoints as well as for composite time-to-first-event endpoints within the context of classical single-stage designs (Chap. 5) and in group-sequential or adaptive designs (Chap. 6). In Chap. 7, exemplary source code written in the software `R` implementing the different approaches introduced in Part II is provided to ease the application in practice.

In Part III, the focus lies on multiple test problems which are of interest if the composite endpoint alone is not sufficient to provide enough information on treatment efficacy and is therefore simultaneously tested along with its individual (main) components to ease the interpretation of the results. Chapter 8 provides a general mathematical introduction on how to derive the correlation between the test statistics of a composite endpoint and an individual component. This correlation structure can be implemented within a multiple test procedure in several ways. As before in Part II, the test hypotheses, the test statistics, as well as the sample size calculation algorithms are provided for classical single-stage designs in Chap. 9 and for group-sequential or adaptive designs in Chap. 10. Again, Part III concludes with a Chap. 11 providing `R` code to implement the different methods.

A completely different approach to ease the interpretation of a composite endpoint without formulating a multiple test problem is to directly define a weighted composite effect measure, where the weights reflect the clinical relevance of the different components. Part IV presents weighted effect measures for composite binary endpoints in Chap. 12 and for composite time-to-first-event endpoints in Chap. 13. Moreover, alternative weighting strategies which are prominently discussed in the statistical and medical literature are critically reviewed in Chap. 14. As in the previous parts, Part IV concludes with a Chap. 15 providing the related R code of the different methodologies.

Whereas Parts II–IV are dedicated to the formulation of an adequate test strategy for the confirmatory efficacy proof based on the composite endpoint, the aim of Part V is to address the issue of additionally evaluating the individual components which is a standard guideline recommendation. In Chap. 16, several commonly met descriptive methods to assess the impact of the treatment under investigation on the individual components are discussed. In contrast, Chap. 17 investigates simple confirmatory analysis strategies to potentially obtain additional confirmatory evidence for the components even if the underlying multiple test problem does not correspond to the formal efficacy claim for which the trial is powered. As before, Chap. 18 provides the corresponding R code of the discussed methods.

Finally, Part VI is dedicated to illustrate all methods presented within this book by means of real clinical trial scenarios. As for a specific clinical trial the definition of an adequate planning and analysis strategy requires implementation of several aspects and methods discussed within this book, we decided to provide an entire exemplary part at the end of the book instead of illustrating each method separately. Moreover, there often exist several alternative planning or analysis approaches to address the trial-specific challenges which should be compared and outweighed against each other. We therefore decided to recall the exemplary clinical trials first introduced in Chap. 4 in Part I and to present different planning and analysis strategies for each of them subsequently. By this, the different statistical approaches along with their advantages and challenges can be directly compared. Part VI is divided into Chap. 19 describing clinical trial scenarios for (composite) binary endpoints and Chap. 20 addressing (composite) time-to-first-event endpoints.

In conclusion, this book gives a comprehensive overview on all important issues on how to plan and evaluate clinical trials with a composite primary endpoint to assure the choice of proper and efficient methods as well as a clinically meaningful and valid interpretation of the results. The book gives practical advice for statisticians and for medical experts involved in the planning and analysis of clinical trials. For readers from the mathematical field, we also provide the underlying statistical theory in order to give a sound mathematical background. For readers which are mainly interested in the application of the methods, we illustrate all approaches with real clinical trial examples and moreover provide the required software code for a fast and easy implementation. The book also discusses all presented methods in the context of relevant guidelines related to the topic. Therefore, the book addresses many issues which are relevant for biostatisticians and medical experts involved in clinical research.

Berlin, Germany
Heidelberg, Germany
Heidelberg, Germany
February 2018

Geraldine Rauch
Svenja Schüler
Meinhard Kieser

Contents

Part I
General Introduction to Composite Endpoints

Clinical trials aim at demonstrating the efficacy of a new treatment or intervention. In phase III drug trials, marketing approval is based on a significant and relevant treatment effect in the primary endpoint. Some interventions, however, never reach phase III, as previous phase I or II studies did not reveal the required effects. From this perspective, the choice of a meaningful and relevant primary endpoint is a key issue when planning a trial.

Composite endpoints combine several variables of clinical interest within a single outcome. They are most often defined as event variables or as time-to-first-event variables. In particular in the fields of cardiology and oncology, the primary variable is commonly defined as a composite endpoint. In these areas, clinical relevant events like death often correspond to very rare events resulting in an inadequately high sample size. By incorporating other event types, such as myocardial infarction or tumor progress, the expected number of events is increased with the intent to increase power or reduce sample size, respectively. Moreover, composite endpoints allow to analyze several outcome variables of interest without the need to adjust for multiplicity. However, the interpretation of a composite endpoint can be difficult, as the effect observed for the composite does not necessarily reflects the effects for the components. Indeed, a negative effect in one component can be masked by a high positive effect in another. In the planning stage, it is therefore recommended to combine only endpoints within a composite which are assumed to be influenced in the same direction. However, there exist a number of study examples which demonstrate that assumptions made in the planning stage turned out to be wrong.

In this part of the book, a general definition of composite endpoint will be provided along with the rationale for their use and a description of related challenges.

Chapter 1
Definition and Rationale

In clinical trials, the primary efficacy variable often corresponds to a so-called composite endpoint. Such an endpoint combines information from different variables within a single net measure. The choice of a composite primary endpoint can be motivated by different considerations. In order to understand and discuss the rationale for the use of composite endpoints and the related problems, we will first start by providing an explicit definition of composite endpoints. Once the definition is clear, the rationale for the use of composite endpoints is easily deduced.

1.1 Definitions and Types of Composite Endpoints

A composite endpoint is generally defined as an outcome combining several endpoints of interest within a single variable. These constituting endpoints are referred to as the components of the composite. In the literature, several types of composite endpoints are considered (Kleist 2008; Chi 2005; CPMP 2002), where the differences lie in the underlying scale levels of the components. In the following, the different types of composite endpoints will be defined and discussed.

1.1.1 Composite Binary Event Endpoints

When the components are given by binary event indicators, a composite endpoint can be defined as an "all-type-event indicator" which equals 1 whenever at least one of the component indicators is 1 and equals 0 when all component indicators equal 0. Such a composite event indicator will be referred to as a "composite binary endpoint" throughout this book. The components of a composite binary event endpoint correspond to several binary event types of clinical interest, which have to be specified in advance.

Generally, composite binary event endpoints are applicable if all events can be observed within a relatively short time frame provided that loss to follow-up is

G. Rauch et al., *Planning and Analyzing Clinical Trials with Composite Endpoints*, Springer Series in Pharmaceutical Statistics,
https://doi.org/10.1007/978-3-319-73770-6_1

not an issue. Moreover, the follow-up duration has to be equal for all patients. To illustrate the concept, an example is given by Yuksel et al. (2010) who assessed the effect of a community pharmacist screening program intending to improve patient's activity concerning diagnosis and treatment of osteoporosis. The primary endpoint in this study was defined as a composite of two indicator variables. The first indicator equals 1 if a new initiation of osteoporosis medication is given, and the second equals 1 whenever a bone mineral density test was performed. Another example is given by an active-controlled clinical trial on renal transplanted patients (Kahan 2000), where the primary endpoint was "treatment failure," which was defined as occurrence of "biopsy-confirmed acute rejection episode," "graft loss," or "death within 6 months after transplantation." Other clinical trial examples based on composite binary event endpoints are provided in Chap. 4. Different planning and analysis strategies for these exemplary clinical trials are provided in Part VI of this book.

Composite binary endpoints are usually evaluated by statistical tests for cross-tables, e.g., by the well-known chi-square test to compare independent rates or by means of the equivalent normal approximation test for rates. For covariate adjustment, logistic regression models can be applied.

It should be noted that clinical response variables can often also be interpreted as composite binary event endpoints. A treatment response is thereby given by a binary indicator, where the response indicator equals 1 if several binary response criteria are met simultaneously. Note that although this is different from the definition above where it was stated that the composite equals 1 if *at least one* component indicator equals 1, the difference simply lies in an inverse coding of the underlying indicators. Response variables are commonly defined as secondary endpoints whose results are of particular relevance in case of small but statistically significant improvements in the primary variables; compare, for example, the Committee for Proprietary Medicinal Products (CPMP) guideline "Points to Consider on Multiplicity Issues in Clinical Trials" (CPMP 2002). As an example, in the field of Alzheimer's disease, the "Guideline on Medical Products for the Treatment of Alzheimer's Disease and Other Dementias" (CHMP 2008) encourages to perform a responder analysis after evaluation of the primary endpoints, where nonresponders are defined as patients with no relevant improvement in cognition or a worsening in certain other domains. Of course, there also exist situations where the response variable is the primary endpoint, for example, in the field of Parkinson's disease (CHMP 2012).

1.1.2 Composite Time-to-First-Event Endpoints

When the components are given by event times, a composite can be defined as a time-to-first-event variable, where again the event is one out of several predefined event types of clinical interest. In clinical study applications and in the statistical literature, the expression "composite endpoints" usually is reserved for a time-to-first-event variable. Thus, this type of composite endpoints refers to the most relevant and most important one. This is due to the fact that time-to-event endpoints generally play a major role in clinical trial applications. Thereby, the patients are observed until either the event of interest occurs or until the study is stopped. Patients without experiencing an event until the end of the study or patients who are lost to follow-up during the observational period are treated as censored observations. It is important to incorporate censored observations when comparing event times between groups, as patients leaving the study before experiencing an event reduce the set of patients at risk. The application of composite time-to-first-event endpoints can often be found in the field of oncology, where common primary outcome variables are given by "progression-free survival" or "metastasis-free survival," which both can be interpreted as a composite endpoint with the components "death" and "disease progression" or "occurrence of metastasis," respectively. Examples for clinical trials with composite time-to-first-event variables are provided in Chap. 4.

Composite time-to-first-event endpoints are commonly evaluated by survival analysis techniques, where treatment comparison is usually performed by the standard logrank test or a Cox-model to adjust for covariates. It should be noted, however, that the components forming a composite time-to-first-event endpoint usually define competing risks (Beyersmann et al. 2012; Kay and Schumacher 1982). The competing risk scenario will be further discussed in Sect. 2.3.

1.1.3 A Note on Clinical Scores

In general, it is possible to combine endpoints of any scale level within a combined outcome variable. However, in the literature the expression "composite endpoints" is usually reserved for the combination of either binary event endpoints or event time variables. Therefore, this book focuses only on these types of composite endpoints. Nevertheless, there also exist similar concepts for the combination of continuous or ordinal variables or even for the combination of differently scaled variables. Many clinical scores are defined as a combined measure of several rating scales assessing different aspects of the outcome variable of interest. An example is given by the Hamilton Rating Scale for Depression (HAM-D) which is used in depression studies (CHMP 2013; Behnke et al. 2003). Although the rationale and the problems related to the use of scores are similar to those for composite endpoints, they are not further discussed here, as the differences in the underlying statistical methodology would go beyond the scope of this book.

1.2 Rationale for the Use of Composite Endpoints

Both types of composite endpoints, binary and time-to-first-event, have in common that they combine different event types of clinical interest. This motivates the rationale for the use of composite endpoints as follows.

1.2.1 Augmenting Power

When planning a trial with a binary or a time-to-event endpoint, the power generally sensitively depends on the observed total number of events. This can become a problem if the outcome of interest corresponds to a relatively rare event (Ferreira-Gonzàles et al. 2007). For example, patients experiencing cardiovascular events in the past usually have a good overall survival prognosis making it difficult to assess differences in the survival curves between several treatments; compare the OMEGA Trial (Rauch et al. 2010) presented in Chap. 4 as a clinical trial example. Using a rare event as primary outcome may result in a low power or, equivalently, in the need for a large sample size and a high time and financial effort which is often not feasible. By combining several types of events into a so-called composite endpoint, the number of expected events is increased with the intention to increase the overall treatment effect and thereby increasing the power. By this, the required sample size and the study duration can be reduced. Often a composite endpoint is the only solution to make a trial feasible at all. Thus, the standard situation for the use of composite endpoints occurs when the events of interest are rare; compare also Cannon (1997).

1.2.2 Avoiding Multiplicity

Another possible motivation for the use of a composite endpoint is that the effect of a new intervention can only be adequately described by several event variables of interest. Instead of formulating a multiple test problem for several primary endpoints, which usually comes along with a loss in power or an increase in sample size, the ICH E9 guideline "Statistical Principles for Clinical Trials" (ICH 1998) states that a composite outcome

> [...] addresses the multiplicity problem without requiring adjustment to the type I error.

For example, Lubsen and Kirwan (2002) argue that atherosclerosis can result in a variety of clinical manifestations and complications, so that treatment of atherosclerosis cannot be evaluated by a single outcome variable. Also, efficacy of HIV treatment is usually evaluated by using several surrogate endpoints believed to be of prognostic relevance (Wittkop et al. 2010). In such cases, a composite endpoint can be used to assess a clinical effect of interest which is not directly related to a single specific event variable but refers to several event types.

References

Behnke, K., Søgaard, J., Martin, S., Bäuml, J., Ravindran, A. V., Ågren, H. et al. (2003). Mirtazapine orally disintegrating tablet versus Sertraline: A prospective onset of action study. *Journal of Clinical Psychopharmacology, 23*, 358–364.

Beyersmann, J., Allignol, A., & Schumacher, M. (2012). *Competing risks and multistate models with R*. New York: Springer.

Cannon, C. P. (1997). Clinical perspectives on the use of composite endpoints. *Controlled Clinical Trials, 18*, 517–529.

Chi, G. Y. H. (2005). Some issues with composite endpoints in clinical trials. *Fundamental & Clinical Pharmacology, 19*, 609–619.

CHMP. (2008). *Guideline on medical products for the treatment of Alzheimer's disease and other dementias*. CPMP/EWP/553/95. http://www.ema.europa.eu/docs/en_GB/document_library/Scientific_guideline/2009/09/WC500003562.pdf. Accessed February 6, 2018.

CHMP. (2012). *Guideline on clinical investigation of medical products in the treatment of Parkinson's disease*. EMA/CHMP/330418/2012. http://www.ema.europa.eu/docs/en_GB/document_library/Scientific_guideline/2012/07/WC500129601.pdf. Accessed February 6, 2018.

CHMP. (2013). *Guideline on clinical investigation of medicinal products in the treatment of depression*. EMA/CHMP/185423/2010. http://www.ema.europa.eu/docs/en_GB/document_library/Scientific_guideline/2013/05/WC500143770.pdf. Accessed February 6, 2018.

CPMP. (2002). *Points to consider on multiplicity issues in clinical trials*. CPMP/EWP/908/99. http://www.ema.europa.eu/docs/en_GB/document_library/Scientific_guideline/2009/09/WC500003640.pdf. Accessed February 6, 2018.

Ferreira-Gonzàles, I., Permanyer-Miralda, G., Domingo-Salvany, A., Busse, J. W., Heels-Ansdell, D., Montoi, V. M., et al. (2007). Problems with the use of composite end points in cardiovascular trials: Systematic review of randomised controlled trials. *British Medical Journal, 334*, 786–793.

ICH. (1998). *Statistical principles for clinical trials - E9*. ICH. http://www.ich.org/fileadmin/Public_Web_Site/ICH_Products/Guidelines/Efficacy/E9/Step4/E9_Guideline.pdf. Accessed February 6, 2018.

Kahan, B. D. (2000). Efficacy of Sirolimus compared with Azathioprine for reduction of acute renal allograft rejection: A randomised multicentre study. *The Lancet, 356*, 194–202.

Kay, R., & Schumacher, M. (1982). Unbiased assessment of treatment effects on disease recurrence and survival in clinical trials. *Statistics in Medicine, 2*, 127–161.

Kleist, P. (2008). Klinische Studien mit zusammengesetzten Endpunkten (Composite Endpoints) - Ihre Durchführung und Interpretation. *Schweizerisches Medizinisches Forum, 47*, 908–912.

Lubsen, J., & Kirwan, B.-A. (2002). Combined endpoints: Can we use them? *Statistics in Medicine, 21*, 2959–2970.

Rauch, B., Schiele, R., Schneider, S., Diller, F., Victor, N., Gohlke, H., et al. (2010). OMEGA, a randomized, placebo-controlled trial to test the effect of highly purified omega-3 fatty acids on top of modern guideline-adjusted therapy after myocardial infarction. *Circulation, 122*, 2152–2159.

Wittkop, L., Smith, C., Fox, Z., Sabin, C., Richert, L., Aboulker, J. P., et al. (2010). Methodological issues in the use of composite endpoints in clinical trials: Example from the HIV field. *Clinical Trials, 7*, 19–35.

Yuksel, N., Majumdar, S. R., Biggs, C., & Tsuyuki, R. T. (2010). Community pharmacist-initiated screening program for osteoporosis: Randomized controlled trial. *Osteoporosis International, 21*, 391–398.

Chapter 2
Challenges of Composite Endpoints

In the previous chapter, it has been shown that a composite endpoint can help to increase the power or save sample size and can define an alternative to a multiple endpoint test problem. Despite these favorable properties of composite endpoints, there also exist some major problems and challenges related to composite endpoints which will be outlined in the following.

2.1 Uncertainties in the Planning Stage

A major difficulty in the planning stage of a clinical trial with a composite primary endpoint is that the sample size calculation is based on much more parameter assumptions as compared to a clinical trial with a single event endpoint. This is due to the fact that the assumed effect for the composite endpoint, which is used for sample size calculation, depends on the effects in the single components and on the correlation between them. Therefore, the required sample size for a clinical trial with a composite endpoint is usually based on a low level of evidence. As a consequence, the trials are often under- or overpowered. The calculation of the required sample size for such studies is thus generally a challenging task.

2.2 Interpretation of Results

With respect to the analysis and the interpretation of clinical trials with composite endpoints, another difficulty is that the effect for the composite endpoint does not necessarily reflects the effects for the individual components (Bethel et al. 2008; Freemantle and Calvert 2007; Montori et al. 2005; Freemantle et al. 2003). Even if a statistically significant and clinically relevant effect in the composite endpoint has been observed, it may happen that the effects for some components are of very different magnitude or even point in the opposite direction; see Fig. 2.1 for a schematic illustration.

G. Rauch et al., *Planning and Analyzing Clinical Trials with Composite Endpoints*, Springer Series in Pharmaceutical Statistics,
https://doi.org/10.1007/978-3-319-73770-6_2

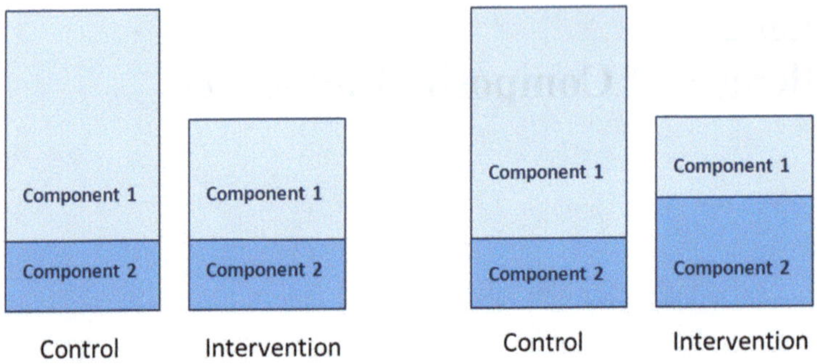

Fig. 2.1 Schematic illustration of the composite effect. Left: composite effect based exclusively on the first component. Right: large effect in the first component masks adverse effect in the second component

If the efficacy of the treatment is judged based on the composite effect alone, this can result in serious misinterpretations. This is especially a problem when the composite consists of "hard" and "soft" endpoints and the soft endpoints refer to the larger effect sizes. For example, the DREAM Trial (The DREAM Trial Investigators 2006) investigated whether rosiglitazone was effective in preventing the development of type 2 diabetes in high-risk patients. The primary endpoint was defined as a composite endpoint of "death" and "newly developed diabetes." The two components are of quite different clinical relevance, and the outcome of the trial was mainly driven by the diabetes endpoint. The death rates, however, were similar in both groups.

2.3 Competing Risks as a Source of Bias

For composite time-to-first-event endpoints, the individual components necessarily correspond to competing risks. The different types of events compete with each other as only one event can be the *first*. In the presence of competing risks, the event rate for a specific event type, which is correctly estimated by the empirical cumulative incidence function, does not only depend on the number of events of this specific type but also on all competing events (Beyersmann et al. 2012), compare Chap. 16. Therefore, the event probabilities cannot be reasonably interpreted without simultaneously looking at all competing event rates. Note that the expressions "event probability" and "event rate" are used interchangeably within this book. For illustration, assume that a novel therapeutic intervention in patients with cardiovascular disease is associated with a 1-year mortality of 0.2 as compared to a death rate of 0.4 in the control group. However, within the same cohort, the rate for nonfatal myocardial infarction could be 0.3 in the treatment group but only 0.1 in the control. In this case, it is apparent that looking exclusively at the component "nonfatal myocardial infarction" would lead to the potentially wrong conclusion that the control might be superior to the treatment group with respect to this type of event. In contrast, the lower event rate for nonfatal myocardial infarction in the control group could exclusively be due to the fact that many patients had died before experiencing a myocardial infarction.

In a multiple test scenario, it is necessary that the treatment effect of an individual component can be judged independently of the treatment effects of the other components. In this context, it is therefore recommended to analyze the treatment effects based on statistical tests for the cause-specific hazards (logrank test or Cox-model), which are not directly influenced by competing events, as these are treated as censored observations (Rauch et al. 2014; Rauch and Beyersmann 2013). Nevertheless, the estimates for the cumulative incidence functions remain useful, because in clinical practice the event probabilities are easier to communicate than hazard ratios and they help to provide a good understanding of the competing risk scenario.

2.4 Follow-Up Beyond the First Event

Composite time-to-first-event variables only take the *first* occurring event into account. This does not necessarily imply that there exist no further events but only that subsequent events are not investigated. Chi (2005) recommends to continue follow-up after the first (nonfatal) event and to take this information into account when analyzing the individual components.

In a descriptive presentation of the individual components, this recommendation could be followed by presenting the event frequencies for each component. However, comparing absolute or relative event frequencies between groups is challenging due to the underlying competing risk scenario; compare Part V of this book. A confirmatory analysis of recurrent events is difficult, as well. For illustration, consider the composite time-to-first-event endpoint "progression-free survival" (which is given as time to death or progression). If the aim is to analyze "death" as a single component, all patients with "progression" as first event would be treated as censored observations in a time-to-first-event analysis. A naive confirmatory analysis ignoring events occurring before the event of interest could be given by analyzing "time to death" irrespective of potential previous "progression" events. However, it should be noted that (a) patients dying without a previous "progress" event and (b) patients who have a "progress" before "death" usually do not follow the same survival distribution; compare also Prentice et al. (1978). However, a common survival distribution for all individuals within one group is the standard assumption for the common survival analysis techniques. Therefore, a standard logrank test based on the ignorance of other event types can be biased. In addition, this kind of analysis is impossible in case of a fatal competing event.

More complex recurrent event models exist which are discussed in more detail in Chap. 5 in Part II of this book. However, these models are also based on restrictive assumptions usually not met in practice. A further solution to study recurrent events, also discussed in Chap. 5 of Part II, would be to use more complex multistate models which model all transition hazards between different subsequent even t types (Beyersmann et al. 2012).

To conclude, modeling recurrent events is a challenging task and based on a number of restrictive assumptions. Recommendations to incorporate the information from subsequent events in descriptive or confirmatory analyses are therefore not easily realized. A more detailed discussion on this aspect is provided in Parts II and V of this book.

References

Bethel, M. A., Holman, R., Haffner, S. M., Califf, R. M., Huntsman-Labed, A., Hua, T. A. et al. (2008). Determining the most appropriate components for a composite clinical trial outcome. *American Heart Journal, 156*, 633–640.

Beyersmann, J., Allignol, A., & Schumacher, M. (2012). *Competing risks and multistate models with R*. New York: Springer.

Chi, G. Y. H. (2005). Some issues with composite endpoints in clinical trials. *Fundamental & Clinical Pharmacology, 19*, 609–619.

Freemantle, N., & Calvert, M. (2007). Composite and surrogate outcomes in randomised controlled trials. *British Medical Journal, 334*, 756–757.

Freemantle, N., Calvert, M., Wood, J., Eastaugh, J., Griffin, C. (2003). Composite outcomes in randomized trials: Greater precision but with greater uncertainty? *Journal of the American Medical Association, 289*, 2554–2559.

The DREAM (Diabetes REduction Assessment with ramipril and rosiglitazone Medication) Trial Investigators (2006). Effect of Rosiglitazone on the frequency of diabetes in patients with impaired glucose tolerance or impaired fasting glucose: A randomised controlled trial. *The Lancet, 368*, 1096–1105.

Montori, V. M., Permanyer-Miralda, G., Ferreira-González, I., Busse, J. W., Pacheco-Huergo, V., Bryant, D. et al. (2005). Validity of composite end points in clinical trials. *British Medical Journal, 330*, 594.

Prentice, R. L., Kalbfleisch, J. D., Peterson, A.V., Jr., Flournoy, N., Farewell, V. T., & Breslow, N. E. (1978). The analysis of failure times in the presence of competing risks. *Biometrics, 34*, 541–554.

Rauch, G., & Beyersmann, J. (2013). Planning and evaluating clinical trials with composite time-to-first-event endpoints in a competing risk framework. *Statistics in Medicine, 32*, 3595–3608.

Rauch, G., Kieser, M., Ulrich, S., Doherty, P., Rauch, B., Schneider, S. et al. (2014). Competing time-to-event endpoints in cardiology trials: A simulation study to illustrate the importance of an adequate statistical analysis. *European Journal of Preventive Cardiology, 21*, 74–80.

Chapter 3
Guideline View: Recommendations and Requirements

In the previous chapter, the challenges and problems related to composite endpoints have been outlined. A number of guidelines exist discussing composite endpoints in various contexts such as in the context of clinical trial application for drugs or medicinal products or in the context of benefit-risk assessment. It is in the nature of things that the existing guidelines are revised and new ones are developed continuously. Therefore, it is not possible to give an exhaustive overview on the perspective of guidelines related to composite endpoints. Instead, in the following some relevant guideline documents will be highlighted, and major topics related to composite endpoints addressed in these guidelines will be outlined.

3.1 Guidelines Related to Composite Endpoints

In the following, three types of guidelines that deal with composite endpoints are distinguished. The first category concerns guidelines on clinical trial methodology and on the presentation of clinical trials and their results. The most prominent document among these guidelines is the International Conference on Harmonization (ICH) E9 guideline "Statistical Principles for Clinical Trials" (ICH 1998). This guideline covers all statistical aspects of clinical trial design, conduct, analysis, and reporting as well as considerations for overall clinical development. Section 2.2.3 of this document is exclusively dedicated to aspects concerning composite variables. In addition, the document of the Committee for Proprietary Medicinal Products (CPMP) referred as "Points to Consider on Multiplicity Issues in Clinical Trials" addresses specific statistical aspects of clinical trial methodology, namely, that of how to deal with multiple comparisons in regulatory applications (CPMP 2002). An updated and amended draft version "Guideline on Multiplicity Issues in Clinical Trials" of the Committee for Medicinal Products for Human Use (CHMP) was recently published (CHMP 2016). In there, various aspects on how to handle composite variables statistically with respect to regulatory claims are discussed in

G. Rauch et al., *Planning and Analyzing Clinical Trials with Composite Endpoints*, Springer Series in Pharmaceutical Statistics,
https://doi.org/10.1007/978-3-319-73770-6_3

Section 6 of the CPMP guidance (CPMP 2002) and in Section 9 of the updated draft guidance of the CHMP (CHMP 2016). Recently, the American Food and Drug Administration (FDA) presented a draft guidance on "Multiple Endpoints in Clinical Trials" (FDA 2017) which was further commented by Sankoh et al. (2017) and Snapinn (2017). In there, composite endpoints are discussed within Section III.C. While the CHMP document (CHMP 2016) is a general guidance on multiplicity aspects, the FDA draft guidance focuses on issues related to challenges when assessing more than one outcome in a confirmatory analysis (FDA 2017). Moreover in 2013, the FDA implemented a document on "Design Considerations for Pivotal Clinical Investigations for Medical Devices" to accommodate the peculiarities of this type of clinical trials (FDA 2013). In Sect. 7.1 on "Endpoints in Clinical Studies," the topic of composite endpoints is addressed. Moreover, the term is also mentioned in the glossary. The FDA Guidance for Industry on "Clinical Studies Section of Labeling for Human Prescription Drug and Biological Products—Content and Format" assists applicants in deciding which information to include in the clinical studies section prescription drug labeling and how to present it (FDA 2006). Composite endpoints are discussed in Section *III.A* of this document where general principles for the description of trials are given.

The second type of guidelines presented here considers methods for benefit-risk and health technology assessments (HTA). In this context, the following institutions and guidelines deal with issues on composite endpoints: The German Institute for Quality and Efficacy in Health Care (IQWiG) conducts assessments of the advantages and disadvantages of medical procedures. In the "General Methods, Version 4.2" document of the IQWiG, Section 8.1.5 is concerned with the interpretation of composite endpoints (IQWIG 2015). The European network for Health Technology Assessment (EUnetHTA) supports the collaboration between European HTA organizations. Among others, the network develops guidelines and tools for HTAs. The guideline "Endpoints used for Relative Effectiveness Assessment: Composite Endpoints" discusses various aspects of composite endpoints and provides recommendations with respect to construction, analysis, synthesis of evidence, and reporting (EUnetHTA 2015).

The third class of guidelines concerns documents that provide guidance for the evaluation and approval of medicinal products for the prevention or treatment of specific diseases. As examples, we consider the "Guideline on the Evaluation of Medicinal Products for Cardiovascular Disease Prevention" (CHMP 2008a), the "Guideline on Clinical Investigation of Medicinal Products for the Treatment of Acute Heart Failure" (CHMP 2008b) of the Committee for Medicinal Products for Human Use (CHMP), and the FDA Guidance for Industry "Clinical Trial Endpoints for the Approval of Cancer Drugs and Biologics" (FDA 2007).

3.2 Guideline Recommendations

The two main reasons for using composite endpoints that were already outlined in Chap. 1 are also pointed out in the corresponding guideline documents. As already mentioned earlier, the ICH E9 guideline (ICH 1998) states that

> This approach addresses the multiplicity problem without requiring adjustment to the type I error.

and the EUnetHTA guideline on "Endpoints used for Relative Effectiveness Assessment: Composite Endpoints" notes that

> The main advantage of composite endpoints is a gain in statistical efficiency of a trial [...].

and mentions the aspects of reduced sample size, length, and costs (EUnetHTA 2015, p. 5). Similar arguments are given in the most recent FDA draft guideline (FDA 2017).

The recommendations given in guidelines concerning the correct use of composite endpoints can be classified whether they deal with planning, analysis, or with claims that can be deduced from the results. For the planning stage, prespecification and careful choice of the components as well as the method of combining them is stressed in various guideline documents; compare, for example, the related documents of FDA (2017), EUnetHTA (2015), IQWIG (2015), FDA (2013), CHMP (2016), CPMP (2002), and ICH (1998). With respect to the choice of the components, the EUnetHTA guideline (EUnetHTA 2015, p. 7) states that

> [...] components of similar clinical importance and sensitivity to interventions should preferably be combined.

Similar statements can be found in the recent FDA draft guideline (FDA 2017). Correspondingly, the CHMP "Guideline on the Evaluation of Medical Products for Cardiovascular Disease Prevention" (CHMP 2008b, p. 6) notes that

> including in the composite components which have a markedly different weight in term of clinical benefit is discouraged

and further points out

> The more clearly components of a composite endpoint directly refer to the disease process, the less there is any problem of interpretation.

An example where violation of the latter principle leads to exclusion of a composite variable as primary endpoint is given in the FDA Guidance for Industry "Clinical Trial Endpoints for the Approval of Cancer Drugs and Biologics." Here, the time-to-treatment failure (TTF) is defined as a composite endpoint measuring the time from randomization to discontinuation of treatment for any reason. This may, for example, be due to disease progression, toxicity of the treatment, or death from any reason. This combination leads to a mix of disparate aspects and overall to an unspecific endpoint. Consequently, the guideline makes a clear statement (FDA 2007, p. 9)

> TTF is not recommended as a regulatory endpoint for drug approval. TTF does not adequately distinguish efficacy from these additional variables.

Another important aspect with respect to the choice of the components is the expectation that they show "directional concordance"; compare CHMP (2008b, p. 8). Accordingly, the CPMP "Points to Consider on Multiplicity Issues" (CPMP 2002, p. 8) recommends

> [...] to include only components for which it can be assumed that treatment will influence them similarly.

This statement is slightly corrected by the new draft document of the CHMP where it says (CHMP 2016, p. 13)

> A composite endpoint must make sense from a clinical perspective. For any component that is included in the composite, it is usually appropriate that any additional component reflecting a worse clinical event is also included. [...] Excluding such events, with an argument that no beneficial effect can be expected [...] introduces difficulties for analysis and interpretation that should be approached carefully.

The later recommendation takes account of the dilemma that the clinically most meaningful components and the components with a similar effect do not necessarily coincide. The assessment of whether the assumption of similar component effects really holds true is echoed in recommendations for the analysis of clinical trials with composite endpoints that we address in the following paragraph.

For the analysis step, guidelines unanimously emphasize that not only the composite endpoint but additionally the components are to be analyzed separately; compare the related documents EUnetHTA (2015), IQWIG (2015), CHMP (2008a), FDA (2006), CPMP (2002), and ICH (1998). In this context, the CPMP "Points to Consider on Multiplicity Issues" especially addresses the concern that an adverse effect of the treatment on one or more components may be masked by a positive overall effect in the composite endpoint. Analogously, the EUnetHTA guideline (EUnetHTA 2015, pp. 7–8) states

> It is recommended to check that clinically important components of the composite endpoints are not affected negatively by the treatment, as some treatments may have negative effect on one component which can be masked by a large beneficial effect of the remaining components.

The CPMP "Points to Consider on Multiplicity Issues" and the updated draft version (CHMP 2016) explicitly claim demonstration that the clinically more important components should at least not be affected negatively, and it is recommended to define the strategy for the related assessment already in the planning stage (CHMP 2016, p. 14; CPMP 2002, pp. 9–10). The CHMP "Guideline on the Evaluation of Medicinal Products for Cardiovascular Disease Prevention" (CHMP 2008a, p. 6) mentions

> Any point estimate considerably in favor of the comparator is a matter of concern.

In the same spirit, the FDA Guidance on "Design Considerations for Pivotal Clinical Investigations for Medical Devices" (FDA 2013, pp. 25–26) states that

> FDA will also evaluate the effect of the device on each of the component endpoints so that domination of the composite by any of its components or lack of consistency in individual component results can be assessed.

Similar recommendations are given in the recent FDA draft guidance (FDA 2017).

An important aspect for applicants is the claim that can be made based on the trial results. The CPMP "Points to Consider on Multiplicity Issues" (CPMP 2002, p. 8) presents difficulties that may occur when using a composite endpoint as primary variable and emphasizes that

> [...] any effect of the treatment in one of the components that is to be reflected in the indication should be clearly supported by the data.

More specifically, the CHMP "Guideline on the Evaluation of Medicinal Products for Cardiovascular Disease Prevention" requires application of a valid multiple test procedure in order to be allowed to make claims based on components of a composite endpoint (CHMP 2008a, p. 6)

> Any secondary outcome measures on which a claim is to be made should be imbedded into the confirmatory testing strategy appropriately controlling type I error.

The following chapters present methods to properly address these recommendations as well as further challenges occurring in the context of clinical trials with composite endpoints.

3.3 Beyond the Guidelines: Open Issues

The main recommendations from the related guidelines presented in detail above concerning the definition of a valid composite endpoint can be summarized as follows:

1. All components combined in a composite endpoint must be of *clinical relevance* for the objective of the trial. Combining endpoints which are only of small relevance may result in apparently positive treatment results which, however, do not reflect substantial benefit for the patients.

2a. Ideally, the effect sizes for the constituting components are of *similar magnitude* so that each endpoint contributes similar amounts to the effect size of the composite.
or

2b. Each component should be sufficient to prove treatment efficacy on its own. If only one component shows a high positive effect, this should suffice to validate the efficacy of the treatment. In other words, this means that all components should be of *similar clinical relevance*.

Unfortunately, the above requirements are usually hard or even unrealistic to fulfill in clinical practice. For example, in trials evaluating patients with heart disease, most often "death" defines one of the components which clearly is more relevant for the patient than any other possible event of interest. Moreover, the components are usually affected in different amounts by a new intervention, so that neither recommendation 2a nor 2b is fulfilled. In addition, valid estimates of the effect sizes of all components are often not available in the planning stage of a clinical trial, so that the verification of recommendation 2a is usually difficult. The CAPRICORN Trial (The CAPRICORN Investigators 2001), presented in detail in Chap. 4, is a prominent example demonstrating that planning assumptions can be heavily wrong. The original primary endpoint for this trial was "all-cause mortality." During a masked interim analysis, it was noted that overall mortality was lower than anticipated and the primary endpoint was changed to a composite endpoint of "all-cause mortality" or "hospital admission for cardiovascular problems." The final study results, however, showed a significant effect for the mortality endpoint but no significance for the composite endpoint.

If the recommendations given above cannot (all) be met, the interpretation of the composite endpoint becomes difficult if no information on the effects of the single components is taken into account. As a consequence, in addition to the analysis of the composite endpoint, an evaluation of the individual components is a general recommendation from the related guidelines. Therefore, it is of major interest to develop methods which allow a sound planning, analysis, and interpretation of clinical trials with composite endpoints. Related methods will be presented in the following parts of this book.

References

CHMP. (2008a). *Guideline on the evaluation of medicinal products for cardiovascular disease prevention*. EMEA/CHMP/EWP/311890/2007. http://www.ema.europa.eu/docs/en_GB/document_library/Scientific_guideline/2009/09/WC500003290.pdf. Accessed April 9, 2018.

CHMP. (2008b). *Guideline on clinical investigation of medicinal products for the treatment of acute heart failure*. CPMP/EWP/2986/03. http://www.ema.europa.eu/docs/en_GB/document_library/Scientific_guideline/2015/06/WC500187797.pdf. Accessed April 9, 2018.

CHMP. (2016). *Guideline on multiplicity issues in clinical trials*. Draft. EMA/CHMP/44762/2017. http://www.ema.europa.eu/docs/en_GB/document_library/Scientific_guideline/2017/03/WC500224998.pdf. Accessed April 9, 2018.

CPMP. (2002). *Points to consider on multiplicity issues in clinical trials*. CPMP/EWP/908/99. http://www.ema.europa.eu/docs/en_GB/document_library/Scientific_guideline/2009/09/WC500003640.pdf. Accessed April 9, 2018.

The CAPRICORN Investigators (2001). Effect of Carvedilol on outcome after myocardial infarction in patients with left-ventricular dysfunction: The CAPRICORN randomised trial. *The Lancet, 357*, 1385–1390.

EUnetHTA. (2015). *Endpoints used for relative effectiveness assessment: Composite endpoints*. Adapted version (2015) based on endpoints used for relative effectiveness assessment of pharmaceuticals: Composite endpoints - February 2013. EUnetHTA. http://www.eunethta.eu/sites/default/files/sites/5026.fedimbo.belgium.be/files/Endpoints%20used%20for%20Relative%20Effectiveness%20Assessment%20Composite%20endpoints_Amended%20JA1%20Guideline_Final%20Nov%202015_0.pdf. Accessed April 9, 2018.

FDA. (2006). *Guidance for industry - Clinical studies section of labeling for human prescription drug and biological products - Content and format*. FDA. http://www.fda.gov/downloads/RegulatoryInformation/Guidances/ucm127534.pdf. Accessed April 9, 2018.

FDA. (2007). *Guidance for industry - Clinical trial endpoints for the approval of cancer drugs and biologics*. FDA. http://www.fda.gov/downloads/Drugs/.../Guidances/ucm071590.pdf. Accessed April 9, 2018.

FDA. (2013). *Guidance for industry - Design considerations for pivotal clinical investigations for medical devices*. FDA. http://www.fda.gov/downloads/MedicalDevices/DeviceRegulationandGuidance/GuidanceDocuments/UCM373766.pdf. Accessed April 9, 2018.

FDA. (2017). *Guidance for industry - Multiple endpoints in clinical trials*. Draft. FDA. https://www.fda.gov/downloads/Drugs/GuidanceComplianceRegulatoryInformation/Guidances/UCM536750.pdf. Accessed April 9, 2018.

ICH. (1998). *Statistical principles for clinical trials - E9*. ICH. http://www.ich.org/fileadmin/Public_Web_Site/ICH_Products/Guidelines/Efficacy/E9/Step4/E9_Guideline.pdf. Accessed April 9, 2018.

IQWIG. (2015). *General methods, version 4.2*. IQWIG. https://www.iqwig.de/en/methods/methods-paper/general-methods-previous-versions.3021.html. Accessed April 9, 2018.

Sankoh, A. J., Li, H., & D'Agostino, R. B. (2017). Composite and multicomponent end points in clinical trials. *Statistics in Medicine*, *36*, 4437–4440.

Snapinn, S. (2017). Some remaining challenges regarding multiple endpoints in clinical trials. *Statistics in Medicine*, *36*, 4441–4445.

Chapter 4
Clinical Trial Examples

In this chapter, several clinical trial examples will be presented which will be used in the remainder of the book to discuss different clinical trial situations in which composite endpoints might be applied and to illustrate possible planning and analysis strategies for studies with composite endpoints.

4.1 The Osteoporosis Trial

In a randomized, controlled clinical trial on a screening program for osteoporosis (Yuksel et al. 2010), the effect of a community pharmacist screening program on testing and treating osteoporosis was assessed. Patients in the intervention group received an intensive education program on aspects of osteoporosis; patients in the control group received usual osteoporosis care in the community pharmacy. The primary endpoint was defined as a composite binary endpoint of two indicator variables. The first indicator equals 1 if a new initiation of osteoporosis medication was done; the second equals 1 whenever a bone mineral density test was performed. Note that in contrast to the other examples presented within this book, a higher event rate corresponds to a more favorable outcome in this application. The planned total sample size of 262 patients provides a power of 0.95 to detect an increase of the composite success rate from 0.2 in the usual care group to 0.4 in the intervention group for a one-sided significance level of 0.025 assuming equal group allocation.

The study showed a clear superiority of the intervention with respect to the composite (success rate of 0.22 observed in the intervention group versus 0.11 in the control group, $p < 0.0001$ when tested with the normal approximation test for rates). However, this outcome was mainly driven by the bone mineral density test endpoint (success rate 0.22 in the intervention group versus 0.10 in the control group), whereas the new medication rates were rather small (success rate of 0.05 in the intervention versus 0.02 in the placebo group).

G. Rauch et al., *Planning and Analyzing Clinical Trials with Composite Endpoints*, Springer Series in Pharmaceutical Statistics,
https://doi.org/10.1007/978-3-319-73770-6_4

4.2 The MOMS Trial

In another randomized, controlled study in pregnant women, prenatal repair of myelomeningocele was compared to repair after delivery which corresponds to the standard treatment (Adzick et al. 2011). The primary outcome was defined as a composite binary endpoint of fetal or neonatal death or the need for a cerebrospinal fluid shunt placement among children who survived. The study was planned to detect a decrease in the composite endpoint from 0.85 in the standard treatment group to 0.57 in the intervention group at a one-sided significance level of 0.025 with a power of 0.99. This results in a required total number of 200 patients.

A number of 78 patients was allocated to the prenatal surgery group compared to 80 patients in the postnatal surgery group. The composite endpoint was observed in 53 patients in the prenatal surgery group (corresponding to rate of 0.679) and in 78 patients in the postnatal surgery group (0.975) which correspond to a relevant treatment effect favoring prenatal surgery. However, most of the events were observed for the shunt component (51 events corresponding to an event rate of 0.654 in the prenatal surgery group; 78 events corresponding to an event rate of 0.679 in the postnatal surgery group), whereas for fetal or neonatal death only two events were observed, both in the prenatal surgery group. Thus there is a small but opposite effect for the harmful component death. It can be discussed if this small negative effect is still acceptable or not, but the results illustrate that the interpretation of the composite effect alone can be difficult when the component effects point in adverse directions.

4.3 The OMEGA Trial

The OMEGA Trial was a randomized, placebo-controlled, double-blind, multicenter trial to evaluate the effect of omega-3-acid ethyl esters-90 on time to sudden cardiac death in survivors of acute myocardial infarction (Rauch et al. 2010). The primary endpoint was "time to sudden cardiac death" defined as unexpected death resulting from heart disease, occurring within 1 h of the first symptoms or unwitnessed, overnight. The primary endpoint was thus a time-to-event endpoint but not a composite endpoint. However, "time to death from any cause" was defined as a secondary endpoint, where the different causes of death ("sudden cardiac death," "non-sudden cardiac death," or "non-cardiac death") can be interpreted as different types of event. Patients were randomized in a 1:1 allocation to receive either the intervention or placebo. Recruitment duration was not reported, and the minimal patients' follow-up duration was chosen to be 1 year. The sample size calculation was based on detecting a relative reduction of 0.44 in the 1-year cumulative event rate of sudden cardiac death from 0.035 in the placebo group to 0.019 in the OMEGA group with 0.8 power at a global one-sided significance level of 0.025. Note that for the original sample size calculation, the primary composite endpoint was considered as a binary outcome and not as a time-to-event endpoint, which clearly would have been more appropriate here. Incorporating a drop-out rate of 0.088, the planned total sample size was 3800.

The results of the OMEGA Trial were not provided by means of the corresponding hazard ratios, which would have been the appropriate effect measure, but instead by means of the absolute and relative frequencies thus ignoring the censoring distribution and issues of competing risks. Therefore, the published results should be interpreted with care. However, the published results are sufficient to discuss general problems regarding the choice of the primary endpoint. A total of 1919 patient was allocated to the OMEGA group compared to 1885 in the placebo group. The primary endpoint was reached by exactly the same number of 28 patients in both groups (observed rates 0.015) within the follow-up period of 1 year which correspond to an absence of any treatment effect. However, as the number of observed events and thus the resulting event rates were much smaller than expected in the planning stage, the study was seriously underpowered. This might be a situation where a composite endpoint combining other events of interest with the event "sudden cardiac death" might have been helpful.

4.4 The RENAAL Trial

The RENAAL Trial is a typical example showing how difficult the interpretation of a composite endpoint as primary outcome variable can be. The study was designed as a randomized, double-blind, placebo-controlled clinical trial with one interim analysis to assess whether the angiotensin-II-receptor antagonist losartan shows a therapeutic benefit for nephropathy patients with type 2 diabetes (Brenner et al. 2001, 2000). The primary endpoint was a composite time-to-first-event endpoint where the event types were given as "death," "end-stage renal disease," or "doubling in the baseline serum creatinine concentration." Patients were randomized in a 1:1 allocation to receive either losartan or placebo. Recruitment duration was planned to require 2 years, and the minimal patients' follow-up duration was chosen to be 3.5 years. The sample size calculation was based on detecting a relative reduction of 0.2 in the 5-year cumulative event rate of the composite endpoint from 0.58 in the placebo group to 0.464 in the losartan group with 0.95 power at a global one-sided significance level of 0.025, where the adjusted local significance level of the final analysis was 0.024. Assuming exponentially distributed event times, the above effect corresponds to a hazard ratio of 1.392 (control versus intervention). The total sample size of 1513 patients included a considerable number of patients due to a recruitment overrun.

Although hazard ratios would have been the appropriate effect measures in this context, the results of the RENAAL were provided by means of the absolute and relative frequencies which ignores the problem of censoring and competing risks. As a consequence, the published results should be interpreted with care. Nevertheless, these results are used to discuss general issues with respect to the composite primary endpoint. A number of 975 patients was recruited to the losartan group and 984 patients to the placebo group. The primary composite endpoint was reached by 327 patients in the losartan group (0.435) as compared to 359 patients in the placebo group (0.471) which corresponds to a relevant positive treatment effect. A detailed look into the single component effects reveals a relevantly lower risk in the losartan group for the components "doubling in the baseline serum creatinine concentration" (162 events corresponding to an event rate of 0.216 in the losartan group versus 198 corresponding to a rate of 0.260 in the control group) and "end-stage renal disease" (147 events corresponding to an event rate of 0.196 in the losartan group versus 194 events corresponding to an event rate of 0.255 in the control group), whereas in the particular harmful component "death," a small adverse effect was observed (158 events corresponding to an event rate of 0.210 in the losartan group versus 155 events corresponding to an event rate of 0.203 in the control group). These results illustrate the difficulty in the interpretation of a composite endpoint in case a large effect in a single component masks an adverse effect in another severe component.

4.5 The DREAM Trial

The DREAM Trial was a randomized, double-blind, placebo-controlled study to demonstrate that the drug rosiglitazone can prevent high-risk patients from developing type 2 diabetes (The DREAM Trial Investigators 2006). The primary endpoint of the trial was a time-to-first-event endpoint composed of "death" or "newly developed diabetes." A total of 5269 patients were randomized in a 1:1 allocation ratio. The minimal patients' follow-up duration was chosen to be 2 years. The study was planned to provide a power of 0.9 to detect a relative reduction of 0.22 from a rate of 0.045 at 1 year in the placebo group to a rate of 0.035 in the intervention group at a one-sided significance level of 0.025.

The results showed a clear benefit of the intervention. The all-cause hazard ratio for the composite endpoint was given by 2.5 (control versus intervention). However, the effects of the individual components are of very different magnitude. The hazard ratio for death was 1.10, whereas for newly developed diabetes, it was given by 2.63. Thus, the composite effect is mainly driven by the less severe endpoint "newly developed diabetes" which may question the composite effect as a meaningful measure to assess the treatment effect.

4.6 The CAPRICORN Trial

A very illustrative example from cardiology that shows how difficult it can be to determine an adequate primary endpoint in the planning stage is the CAPRICORN Trial. Therein, Dargie (2000) investigated the long-term efficacy of carvedilol regarding morbidity and mortality in patients with left ventricular dysfunction after acute myocardial infarction. In this multicenter, randomized, double-blind, and placebo-controlled trial, patients were randomly assigned to carvedilol or placebo in a 1:1 allocation. The originally planned primary endpoint was "time to death from any cause". The trial was planned to detect a hazard ratio for death of 1.30 (control versus intervention) with a power of 0.9 for a one-sided significance level of 0.025. Recruitment was planned to continue until 633 target events were observed which was assumed to require 1850 patients in total.

During a masked interim analysis, the data safety monitoring board noted that the overall mortality rate was lower than anticipated, and consequently the study was likely to be seriously underpowered. Therefore it was decided to continue the trial with two primary endpoints of which the first corresponds to the original primary endpoint (all-cause mortality) and the second to a composite time-to-first-event endpoint given as time to death or cardiovascular hospital admission. By adding the composite endpoint, it was intended to increase the number of events and thus augmenting the power of the trial. The new efficacy claim of the trial was to show significance in at least one of the two endpoints with higher preference for the more promising composite endpoint. In terms of the multiple test problem arising from this modified analysis, it was therefore decided to assign a weighted Bonferroni-adjusted local significance level of only 0.005 to the original primary endpoint (all-cause mortality), whereas a much larger local level of 0.045 was assigned to the newly added composite endpoint (time to death or cardiovascular hospital admission). The updated target sample size was based on the new composite time-to-first-event endpoint.

After a mean follow-up of 1.3 years and inclusion of 1959 patients, the target number of events was reached for the new composite endpoint. The results of the CAPRICORN Trial published by the CAPRICORN Investigators (2001) showed a hazard ratio for death of 1.30 (control versus intervention) and a hazard ratio of 1.09 for the new composite endpoint. Neither of the endpoints showed a significant effect. The original primary endpoint "time to death" indeed showed the effect assumed in the planning stage, but the underlying event rate was much smaller leading to a loss in power. The new composite endpoint showed more events, but the treatment effect nevertheless was lower than for the original primary endpoint. The observed results illustrate the problem of the high uncertainty about reasonable treatment effect assumptions in the planning stage. The study design modifications based on the blinded interim results were not appropriate to account for these planning uncertainties. Therefore, it remains an open task to discuss other strategies to overcome incorrect or uncertain planning assumptions.

4.7 The LIFE Trial

The LIFE Trial was designed as a randomized, double-blind, controlled clinical trial to compare the effects of losartan and atenolol on cardiovascular morbidity and mortality in patients with diabetes and hypertension (Dahlöf et al. 2002).

The primary endpoint was a composite time-to-first-event endpoint combining "myocardial infarction," "stroke," and "cardiovascular mortality." The recruitment time was fixed to 2 years and the minimal follow-up duration to 4 years. A total of 9193 patients were allocated to the two treatment groups in a 1:1 ratio. The LIFE study was planned to detect a decrease in the 5-year event rate of 0.15 in the control arm to 0.1275 in the intervention group with a power of 0.8 at a one-sided significance level of 0.025 (Dahlöf et al. 1997). This effect corresponds to a hazard ratio of 1.19 (control versus intervention) when assuming exponentially distributed event times. The final sample size of 9193 patients was calculated for a group-sequential design with two interim analyses and included a certain recruitment overrun.

The results showed a significant and relevant effect in favor of the intervention. The all-cause hazard ratio for the composite endpoint was given by 1.15 (control versus intervention) which is very close to the planning assumption. The individual components of the composite were also analyzed and reported. The observed hazard ratio for cardiovascular mortality was given by 1.12, the hazard ratio for stroke was 1.33, and the hazard ratio for myocardial infarction was given by 0.93 which corresponds to a small adverse effect. Note that the results related to the components are only of descriptive nature and provide no confirmatory evidence. To generate additional confirmatory evidence for the components would therefore have been an interesting option in this application.

References

Adzick, N. S., Thom, E. A., Spong, C. Y., Brock, J. W., Burrows, P. K., Johnson, M. P., et al. (2011). A randomized trial of prenatal versus postnatal repair of myelomeningocele. *New England Journal of Medicine, 364*, 993–1004.

Brenner, M. B., Cooper, M. E., de Zeeuw, D., Grunfeld, J.-P., Keane, W. F., Kurokawa, K., et al. (2000). The losartan renal protection study - Rationale, study design and baseline characteristics of RENAAL (Reduction of endpoints in NIDDM with the angiotensin II antagonist losartan). *Journal of Renin-Angiotensin-Aldosterone System, 1*, 328–335.

Brenner, M. B., Cooper, M. E., de Zeeuw, D., Keane, W. F., Mitch, W. E., Parving, H. H, et al. (2001). Effects of losartan on renal and cardiovascular outcomes in patients with type 2 diabetes and nephropathy. *New England Journal of Medicine, 345*, 861–869.

Dahlöf, B., Devereux, R. B., de Faire, U., Fyhrguist, F., Beevers, G., de Faire, U., et al. (1997). The Losartan intervention for endpoint reduction (LIFE) in hypertension study: Rationale, design, and methods. *American Journal of Hypertension, 10*, 705–713.

Dahlöf, B., Devereux, R. B., Kjeldsen, S. E., Julius, S., Beevers, G., de Faire, U., et al. (2002). Cardiovascular morbidity and mortality in the Losartan Intervention For Endpoint reduction in hypertension study (LIFE): A randomised trial against atenolol. *The Lancet, 359*, 995–1003.

Dargie, H. J. (2000). Design and methodology of the CAPRICORN trial - A randomised double blind placebo controlled study of the impact of carvedilol on morbidity and mortality in patients with left ventricular dysfunction after myocardial infarction. *European Journal of Heart Failure, 2*, 325–332.

The CAPRICORN Investigators (2001). Effect of Carvedilol on outcome after myocardial infarction in patients with left-ventricular dysfunction: The CAPRICORN randomised trial. *The Lancet, 357*, 1385–1390.

The DREAM (Diabetes REduction Assessment with ramipril and rosiglitazone Medication) Trial Investigators (2006). Effect of rosiglitazone on the frequency of diabetes in patients with impaired glucose tolerance or impaired fasting glucose: A randomised controlled trial. *The Lancet, 368*, 1096–1105.

Rauch, B., Schiele, R., Schneider, S., Diller, F., Victor, N., Gohlke, H., et al. (2010). OMEGA, a randomized, placebo-controlled trial to test the effect of highly purified omega-3 fatty acids on top of modern guideline-adjusted therapy after myocardial infarction. *Circulation, 122*, 2152–2159.

Yuksel, N., Majumdar, S. R., Biggs, C., & Tsuyuki, R. T. (2010). Community pharmacist-initiated screening program for osteoporosis: Randomized controlled trial. *Osteoporosis International, 21*, 391–398.

References

[illegible] Dubois ... [illegible] transport ... [illegible]
[illegible]
[illegible]

Part II
Confirmatory Test Problem for a Single (Composite) Endpoint

As discussed in Part I of this book, when planning a clinical trial with the option to use a composite endpoint as primary efficacy variable, the pros and cons have to be carefully outweighed. A composite endpoint usually provides a power advantage but may cause interpretation problems. As an alternative, the most important component of the composite under consideration might as well be considered as the primary endpoint which eases interpretation but may be more challenging with respect to power. Depending on the specific clinical situation at hand, it can thus be preferable to use a single (time-to-)event component as the primary efficacy endpoint. The CAPRICORN Trial described in Sect. 4.6 demonstrates that the choice between a composite primary endpoint and an individual component as primary efficacy measure is not always straightforward. In this part of the book, we therefore introduce the underlying test problem for a clinical trial considering either a composite endpoint or a single component as primary efficacy measure. Thus, the focus lies on a confirmatory test problem for a single endpoint. Thereby, we differentiate between binary endpoints and time-to-event endpoints.

Throughout this book, a controlled clinical trial comparing two treatment options is considered, where the index I denotes the intervention group and the index C is assigned to the control. The composite endpoint CE shall consist of k components $EP_l, l = 1, \ldots, k$. The sample sizes in the intervention and the control group are given by n_I and n_C, respectively. In the first chapter, we consider a standard single-stage design meaning that no interim analyses are performed. Under the correct planning assumptions, single-stage designs are the most efficient in terms of power and required maximal sample size. However, group-sequential and adaptive designs including one or several interim analyses might provide an advantage in power or sample size whenever the planning assumptions are uncertain. Therefore, in the second chapter, we consider group-sequential and adaptive designs.

Chapter 5
The Single-Stage Design

In this chapter, the test problem is introduced for the case of a standard single-stage design, meaning that no interim analysis is performed. The underlying parameters and distributional assumptions, the test problems, and the test statistics will be separately formulated for (composite) binary endpoints and for (composite) time-to-(first-)event endpoints. Moreover, approaches to calculate the required sample size are provided.

5.1 Binary Endpoints

The random variables for the number of patients which experience an event in the component EP_l are denoted by $X^I_{EP_l}$ and $X^C_{EP_l}$, $l = 1, \ldots, k$, respectively. Equivalently, the number of patients with an event in *any* component, that means with an event in the composite endpoint, are given by X^I_{CE} and X^C_{CE}. These random variables follow binomial distributions parametrized by

$$X^I_{CE} \sim Bin(n_I, p^I_{CE}),\ X^C_{CE} \sim Bin(n_C, p^C_{CE}),$$
$$X^I_{EP_l} \sim Bin(n_I, p^I_{EP_l}),\ X^C_{EP_l} \sim Bin(n_C, p^C_{EP_l}),\ l = 1, \ldots, k,$$

where $p^I_{CE}, p^C_{CE}, p^I_{EP_l}, p^C_{EP_l}, l = 1, \ldots, k$, denote the underlying event probabilities. These event probabilities are composed of the probabilities to experience *exclusively* an event of type l denoted as $p^I_{EP_l,ex}$ and $p^C_{EP_l,ex}$, respectively, and of the overlap probabilities to experience an event of type l *and* any other event type or combination of event types in addition. For an arbitrary number of components k, there exist

$$m := \sum_{l=2}^{k} \binom{k}{l} = 2^k - k - 1$$

G. Rauch et al., *Planning and Analyzing Clinical Trials with Composite Endpoints*, Springer Series in Pharmaceutical Statistics,
https://doi.org/10.1007/978-3-319-73770-6_5

disjoint sets of overlap combinations consisting of at least two different event types. The disjoint overlap probabilities for the two groups will be denoted as $p^I_{\cap j}$ and $p^C_{\cap j}$ for $j = 1, \ldots, m$, respectively. It is evident that the event probabilities of the single components are always smaller or equal to the event probabilities of the composite in each group. In general, the event probabilities of the composite are related to the event probabilities of the components as follows

$$p^I_{CE} = \sum_{l=1}^{k} p^I_{EP_l,ex} + \sum_{j=1}^{m} p^I_{\cap j},$$

$$p^C_{CE} = \sum_{l=1}^{k} p^C_{EP_l,ex} + \sum_{j=1}^{m} p^C_{\cap j}.$$

The number of disjoint overlap probabilities for the two groups containing a *specific* component EP_l with $l \in \{1, \ldots, k\}$ is given by

$$m_{EP} := \sum_{l=2}^{k} \binom{k-1}{l}.$$

The corresponding overlap probabilities for the two groups will be denoted as $p^I_{l\cap j}$ and $p^C_{l\cap j}$ for $j = 1, \ldots, m$ and a fixed $l \in \{1, \ldots, k\}$. The event probabilities for the individual components are then given as

$$p^I_{EP_l} = p^I_{EP_l,ex} + \sum_{j=1}^{m_{EP}} p^I_{l\cap j},$$

$$p^C_{EP_l} = p^C_{EP_l,ex} + \sum_{j=1}^{m_{EP}} p^C_{l\cap j}, \; l = 1, \ldots, k.$$

For the special case of mutually exclusive components, which means that $m = 0$, the composite event probabilities can also be written as

$$p^I_{CE} = \sum_{l=1}^{k} p^I_{EP_l,ex} = \sum_{l=1}^{k} p^I_{EP_l},$$

$$p^C_{CE} = \sum_{l=1}^{k} p^C_{EP_l,ex} = \sum_{l=1}^{k} p^C_{EP_l}.$$

Due to the high number of possible overlap combinations, the above formulas for the composite event probabilities and for the individual event probabilities are rather technical. The notation becomes much simpler, when the number of components is fixed to some predefined value. For example, for a composite endpoint consisting of two components, the composite event rates and the event probabilities for the individual components are given as

$$\begin{aligned} p_{CE}^{I} &= p_{EP_1,ex}^{I} + p_{EP_2,ex}^{I} + p_{EP_{1\cap 2}}^{I}, \\ p_{CE}^{C} &= p_{EP_1,ex}^{C} + p_{EP_2,ex}^{C} + p_{EP_{1\cap 2}}^{C}, \\ p_{EP_l}^{I} &= p_{EP_l,ex}^{I} + p_{EP_{1\cap 2}}^{I}, \\ p_{EP_l}^{C} &= p_{EP_l,ex}^{C} + p_{EP_{1\cap 2}}^{C}, \; l = 1, 2, \end{aligned}$$

were $p_{EP_{1\cap 2}}^{I}, p_{EP_{1\cap 2}}^{C}$ denote the overlap probabilities of EP_1 and EP_2 in both groups.

5.1.1 Test Problem

In the common situation that lower event rates correspond to more favorable results, the null hypothesis to be tested in the confirmatory analysis states that the composite event rate in the control group is smaller or equal to the event rate in the intervention group versus the alternative hypothesis that the event rate in the control group is larger than the event rate in the intervention group

$$H_0^{CE} : p_{CE}^{C} - p_{CE}^{I} \leq 0 \text{ versus } H_1^{CE} : p_{CE}^{C} - p_{CE}^{I} > 0. \tag{5.1}$$

The test hypotheses for a single component can be formulated equivalently

$$H_0^{EP_l,\delta_l} : p_{EP_l}^{C} - p_{EP_l}^{I} \leq \delta_l \text{ versus } H_1^{EP_l,\delta_l} : p_{EP_l}^{C} - p_{EP_l}^{I} > \delta_l, \; l = 1, \ldots, k, \tag{5.2}$$

where $\delta_l \leq 0$. If δ_l equals 0, the above test problem corresponds to a superiority test, whereas for $\delta_l < 0$ the component is tested for non-inferiority. Note that if an individual component EP_l is chosen as the unique primary efficacy endpoint, then a superiority test problem is usually preferred, that means $\delta_l = 0$. However, as motivated in Part I of this book, a possible approach to improve the interpretation of the results of clinical trials with composite endpoints is to base the efficacy claim on a multiple test problem including confirmatory tests for the composite and the components. This approach will be formally introduced in Part III of this book. In this latter context, a non-inferiority test problem for the components might be useful. Therefore, the more general formulation of the test hypotheses for an individual component was chosen here.

5.1.2 Test Statistics

The standard approach to test the null hypotheses given above is the normal approximation test for rates, where the test statistics are defined as

$$T_{CE,0} := \frac{\widehat{p}^{C}_{CE} - \widehat{p}^{I}_{CE}}{\widehat{\sigma}_{CE,0}}, \tag{5.3}$$

$$T_{EP_l,\delta_l} := \frac{\widehat{p}^{C}_{EP_l} - \widehat{p}^{I}_{EP_l} - \delta_l}{\widehat{\sigma}_{EP_l,\delta_l}}, \; l = 1,\ldots,k. \tag{5.4}$$

Thereby, the maximum likelihood estimators for the true event rates are given by

$$\widehat{p}^{I}_{CE} := \frac{x^{I}_{CE}}{n_I}, \; \widehat{p}^{C}_{CE} := \frac{x^{C}_{CE}}{n_C},$$

$$\widehat{p}^{I}_{EP_l} := \frac{x^{I}_{EP_l}}{n_I}, \; \widehat{p}^{C}_{EP_l} := \frac{x^{C}_{EP_l}}{n_C}, \; l = 1,\ldots,k,$$

which are the corresponding relative frequencies. For the superiority test problem of the composite endpoint, the restricted maximum likelihood estimator of the treatment effect variance under the null hypothesis restriction $p^{C}_{CE} - p^{I}_{CE} = 0$ is given by

$$\widehat{\sigma}^2_{CE,0} = \widehat{p}_{CE} \cdot (1 - \widehat{p}_{CE}) \cdot \frac{n_I + n_C}{n_I \cdot n_C}, \text{ where } \widehat{p}_{CE} = \frac{\widehat{p}^{I}_{CE} \cdot n_I + \widehat{p}^{C}_{CE} \cdot n_C}{n_I + n_C}. \tag{5.5}$$

As the composite endpoint is usually tested for superiority, the subscript 0 might be omitted, and the notation $T_{CE} = T_{CE,0}$ is used within the remainder of the book.

For the (non-inferiority) test problem of the components, the variance estimator is given by

$$\widehat{\sigma}^2_{EP_l,\delta_l} = \frac{\widetilde{p}^{I}_{EP_l,\delta_l} \cdot (1 - \widetilde{p}^{I}_{EP_l,\delta_l}) + \widetilde{p}^{C}_{EP_l,\delta_l} \cdot (1 - \widetilde{p}^{C}_{EP_l,\delta_l}) \cdot \gamma}{n_I}, \; l = 1,\ldots,k, \tag{5.6}$$

where $\gamma = n_I/n_C$ defines the allocation ratio of sample size and $\widetilde{p}^{I}_{EP_l,\delta_l}$ and $\widetilde{p}^{C}_{EP_l,\delta_l}$ are the restricted maximum likelihood estimators under the null hypothesis restriction $p^{C}_{EP_l} - p^{I}_{EP_l} = \delta_l$, $l = 1,\ldots,k$. Their derivation is described in the Appendix of Farrington and Manning (1990). In there, the estimator $\widetilde{p}^{I}_{EP_l,\delta_l}$ is given as the unique solution within the interval $(-\delta_l; 1)$ of the equation

$$a \cdot x^3 + b \cdot x^2 + c \cdot x + d = 0, \tag{5.7}$$

where

$$a = 1 + \frac{1}{\gamma},$$
$$b = -\left(1 + \frac{1}{\gamma} + \widehat{p}^I_{EP_l} + \frac{1}{\gamma} \cdot \widehat{p}^C_{EP_l} - \delta_l \cdot \left(\frac{1}{\gamma} + 2\right)\right),$$
$$c = \delta_l^2 - \delta_l \cdot \left(2 \cdot \widehat{p}^I_{EP_l} + \frac{1}{\gamma} + 1\right) + \widehat{p}^I_{EP_l} + \frac{1}{\gamma} \cdot \widehat{p}^C_{EP_l},$$
$$d = \widehat{p}^I_{EP_l} \cdot \delta_l \cdot (1 - \delta_l).$$

Defining

$$v = \frac{b^3}{(3a)^3} - \frac{b \cdot c}{6a^2} + \frac{d}{2a},$$
$$u = \operatorname{sgn}(v) \cdot \sqrt{\frac{b^2}{(3a)^2 - \frac{c}{3a}}},$$
$$w = \frac{1}{3} \cdot \left(\pi + a\cos\left(\frac{v}{u^3}\right)\right).$$

The solution of equation (5.7) is given as

$$\widetilde{p}^I_{EP_l,\delta_l} = 2 \cdot u \cdot \cos(w) - \frac{b}{3a}, \tag{5.8}$$
$$\widetilde{p}^C_{EP_l,\delta_l} = \widetilde{p}^I_{EP_l,\delta_l} + \delta_l. \tag{5.9}$$

There exist other methods to estimate the null variance for a non-inferiority test problem; compare, for example, Blackwelder (1982) and Dunnett and Gent (1977). However, as a result of performance comparisons of various methods, the restricted maximum likelihood estimator was recommended by Roebruck and Kühn (1995); compare also Farrington and Manning (1990). Note that for $\delta_l = 0$, the variance estimators (5.5) and (5.6) are equivalent.

The corresponding true variances $\sigma^2_{CE,\Delta_{CE}}$ and $\sigma^2_{EP_l,\Delta_{EP_l}}$ for given treatment effects $\Delta_{CE} := p^C_{CE} - p^I_{CE}$ and $\Delta_l := p^C_{EP_l} - p^I_{EP_l}$, $l = 1, \ldots, k$ are given as

$$\sigma^2_{CE,\Delta_{CE}} = \frac{1}{n_I} p^I_{CE}(1 - p^I_{CE}) + \frac{1}{n_C} p^C_{CE}(1 - p^C_{CE}),$$
$$\sigma^2_{EP_l,\Delta_l} = \frac{1}{n_I} p^I_{EP_l}(1 - p^I_{EP_l}) + \frac{1}{n_C} p^C_{EP_l}(1 - p^C_{EP_l}),\ l = 1, \ldots, k.$$

Using these notations, the test statistics given in (5.3) and (5.4) are approximately normally distributed with

$$T_{CE} \sim N\left(\frac{\Delta_{CE}}{\sigma_{CE,0}}, \frac{\sigma^2_{CE,\Delta_{CE}}}{\sigma^2_{CE,0}}\right) \quad (5.10)$$

$$T_{EP_l,\delta_l} \sim N\left(\frac{\Delta_l - \delta_l}{\sigma_{EP_l,\delta_l}}, \frac{\sigma^2_{EP_l,\Delta_l}}{\sigma^2_{EP_l,\delta_l}}\right), \ l = 1, \ldots, k. \quad (5.11)$$

Under the corresponding null hypotheses (5.1) and (5.2), respectively, the test statistics T_{CE} and T_{EP_l,δ_l} hence asymptotically follow a standard normal distribution. Thus, the corresponding null hypotheses are rejected whenever $T_{CE} \geq z_{1-\alpha}$ or $T_{EP_l,\delta_l} \geq z_{1-\alpha}$, respectively, where α denotes the one-sided significance level.

5.1.3 *Sample Size Calculation*

Clinical trials can only achieve the study aim with sufficient chance if the underlying sample size is large enough. On the contrary, the sample size should be kept as small as possible from an ethical perspective and to save time and financial resources. Therefore, the calculation of the sample size is a key point when planning a clinical trial. The standard approach is to choose the sample size such that the power function reaches a predefined value under prespecified parameter assumptions, which are gained from historical data or are based on clinical experience. If these assumptions are wrongly specified, e.g., as historical data for a valid estimation do not exist or are heterogeneous, the calculated sample size can substantially deviate from the "correct" one. As a consequence, the trial becomes invalid or inefficient imposing ethical and economical concerns. In this section, we will deduce the standard approach for sample size calculation and moreover will present alternatives which guarantee a certain robustness against parameter misspecifications in the planning stage. In addition, in Chap. 6 adaptive and group-sequential designs are introduced which allow to react flexible during the ongoing trial if the observed data deviate from the initial assumptions.

5.1.3.1 Standard Sample Size Calculation

The sample size formula for the superiority test problem formulated in (5.1) can be deduced by identifying the smallest value of the sample size fulfilling the following equation

$$P_{H_1^{CE}}\left(T_{CE} \geq z_{1-\alpha}\right) \geq 1 - \beta. \quad (5.12)$$

It can easily be seen that for a predefined effect $\Delta_{CE} = p_{CE}^C - p_{CE}^I$ under H_1^{CE} solving (5.12) is approximately equivalent to

$$1 - P_{H_1^{CE}}\left(X \le \frac{z_{1-\alpha} \cdot \sigma_{CE,0} - \Delta_{CE}}{\sigma_{CE,\Delta_{CE}}}\right) \ge 1 - \beta, \text{ with } X \sim N(0,1)$$

$$\Leftrightarrow \frac{z_{1-\alpha} \cdot \sigma_{CE,0} - \Delta_{CE}}{\sigma_{CE,\Delta_{CE}}} \le -z_{1-\beta},$$

$$\Leftrightarrow z_{1-\alpha} \cdot \sigma_{CE,0} + z_{1-\beta} \cdot \sigma_{CE,\Delta_{CE}} \le \Delta_{CE},$$

$$\overset{(*)}{\Leftrightarrow} z_{1-\alpha}\sqrt{\left(\frac{\gamma+1}{n_I}\right) p_{CE}^C\left(1-p_{CE}^C\right)} + z_{1-\beta}\sqrt{\frac{p_{CE}^I(1-p_{CE}^I) + \gamma p_{CE}^C(1-p_{CE}^C)}{n_I}} \le \Delta_{CE},$$

$$\Leftrightarrow n_I \ge \left(\frac{z_{1-\alpha}\sqrt{(\gamma+1)\, p_{CE}^C\left(1-p_{CE}^C\right)} + z_{1-\beta}\sqrt{p_{CE}^I(1-p_{CE}^I) + \gamma p_{CE}^C(1-p_{CE}^C)}}{\Delta_{CE}}\right)^2,$$

where $\gamma = n_I/n_C$ defines the allocation ratio of sample size. Thereby, the estimated variance under the null hypothesis was approximated by plugging in the assumed true event rate in the control arm given by p_{CE}^C meaning that

$$\widehat{\sigma}_{CE,0} \approx \sigma_{CE,0} = \left(\frac{\gamma+1}{n_I}\right) p_{CE}^C(1-p_{CE}^C).$$

Note that $(*)$ holds true, as under the null hypothesis H_0^{CE} the event probabilities in both groups are equal, i.e., $p_{CE}^I = p_{CE}^C$. The total required sample size for both groups is given by the smallest integer n_{CE} fulfilling

$$n_{CE} := n_I + n_C \ge \left(1 + \frac{1}{\gamma}\right) \cdot \left(\frac{z_{1-\alpha}\sqrt{(\gamma+1)\, p_{CE}^C\left(1-p_{CE}^C\right)} + z_{1-\beta}\sqrt{p_{CE}^I(1-p_{CE}^I) + \gamma p_{CE}^C(1-p_{CE}^C)}}{\Delta_{CE}}\right)^2. \tag{5.13}$$

The sample size formula for the (non-inferiority) test problem formulated in (5.2) for an individual component EP_l, $l = 1, \ldots, k$ can be deduced equivalently (compare Farrington and Manning (1990)) and is given by

$$n_{EP_l} := n_I + n_C \ge \left(1 + \frac{1}{\gamma}\right) \cdot \left(\frac{z_{1-\alpha}\sqrt{\widetilde{p}_{EP_l}^I(1-\widetilde{p}_{EP_l}^I) + \gamma \widetilde{p}_{EP_l}^C(1-\widetilde{p}_{EP_l}^C)} + z_{1-\beta}\sqrt{p_{EP_l}^I(1-p_{EP_l}^I) + \gamma p_{EP_l}^C(1-p_{EP_l}^C)}}{(\Delta_{EP_l} - \delta_l)}\right)^2, \tag{5.14}$$

where $\widetilde{p}^{I}_{EP_l}, \widetilde{p}^{C}_{EP_l}$ are the assumed event rates under the null hypothesis given in (5.2) and thus $\widetilde{p}^{I}_{EP_l} - \widetilde{p}^{C}_{EP_l} = \delta_l$ for $l = 1, \ldots, k$.

5.1.3.2 Robust Sample Size Calculation

In the standard approach presented above, the sample size of a clinical trial is calculated to achieve the aspired power for a unique fixed parameter constellation under the specific alternative hypothesis and therefore crucially depends on the correctness of these parameter assumptions. An ad hoc method to get a conservative estimate of the required sample size would be to calculate the classical power for a range of different plausible parameter settings and to base the sample size calculation on the worst-case scenario yielding the highest sample size.

Although this is a very easy way to define a sample size which is robust against some misspecifications, this approach might lead to unnecessarily high sample sizes. Another possibility is to replace the fixed parameter assumptions by prior distributional assumptions, which better reflect the uncertainty about these parameters. By integrating the power function with respect to these prior distributions, an expected power can be defined (Rauch and Kieser 2013; Daimon 2008; Spiegelhalter and Keith 2004; Brown et al. 1987). The expected power can be interpreted as a semi-Bayesian power approach. The weights are defined by the choice of the prior distributions which reflect the level of evidence on the parameters. If, for example, there have been several studies in the past supporting specific parameter assumptions but a single study suggests more conservative assumptions, it might be inadequate to base the power calculation exclusively on these worst-case assumption which correspond to a very small level of evidence. Instead, a possible approach could be to give a higher weight on the parameter assumptions from the majority of studies and only a small weight to the single study suggesting the more conservative parameter setting. Shao et al. (2008), Begum and King (2005), and Gillett (1994) discussed the advantages of such an expected power approach compared to the classical frequentist power in different clinical trial situations. For the case of a composite binary endpoint or a single binary event variable, a sample size calculation method based on a predefined value for the expected power was developed as described in the following; compare also Rauch and Kieser (2013).

In a first step, the available knowledge in the planning stage has to be "translated" in adequate prior distribution assumptions for the event rates of the primary endpoint under consideration (either a composite or a single event endpoint). Let the prior distributions be given as

$$p^{C}_{CE} \sim F,\ p^{I}_{CE} \sim G,$$
$$p^{C}_{EP_l} \sim X_l,\ p^{I}_{EP_l} \sim Y_l,\ l = 1, \ldots, k.$$

By integrating the classical frequentist power $P_{H_1^{CE}}(T_{CE} \geq z_{1-\alpha})$ or $P_{H_1^{EP_l,\delta_l}}(T_{EP_l,\delta_l} \geq z_{1-\alpha})$, respectively, with respect to the above priors over all parameter constellations within H_1^{CE} or $H_1^{EP_l,\delta_l}$, respectively, the expected power can be defined as follows

$$Pow_{CE}^{\exp}(F,G) := \iint_{S_{CE}} P_{H_1^{CE}}(T_{CE} \geq z_{1-\alpha}) f(p_{CE}^C) g(p_{CE}^I) dp_{CE}^C dp_{CE}^I, \tag{5.15}$$

$$Pow_{EP_l}^{\exp}(X_l,Y_l) := \iint_{S_{EP_l}} P_{H_1^{EP_l,\delta_l}}(T_{EP_l,\delta_l} \geq z_{1-\alpha})\, x_l(p_{EP_l}^C) y_l(p_{EP_l}^I) dp_{EP_l}^C dp_{EP_l}^I, \quad l = 1,\ldots,k. \tag{5.16}$$

Here the functions $f(*), g(*), x_l(*), y_l(*)$ denote the corresponding densities of the predefined prior distributions, and the integration sets S_{CE}, S_{EP_l} are defined as

$$S_{CE} = \left\{p_{CE}^C \sim F, p_{CE}^I \sim G : p_{CE}^C - p_{CE}^I > 0\right\},$$
$$S_{EP_l} = \left\{p_{EP_l}^C \sim X_l, p_{EP_l}^I \sim Y_l : p_{EP_l}^C - p_{EP_l}^I > \delta_l\right\},\ l = 1,\ldots,k.$$

The expected power can be interpreted as a weighted mean over the frequentist power for different parameter settings, where the weights are determined by the definition of the priors. For one-point prior distributions, the expected power exactly corresponds to the standard frequentist power definition.

The required sample size to reach a predefined level of the expected power must be determined by simulations. A corresponding algorithm is provided in Chap. 7. Note that even if the expected power reaches a desired level for a given sample size, this implies only that the weighted average power is reached. However, the power for a fixed parameter constellation might be lower. It has to be carefully considered in the planning stage if the concept of an average power approach seems acceptable or if it seems more suitable to use the worst-case power approach coming along with a larger sample size.

5.2 Time-to-Event Endpoints Under Proportional Hazards

A composite time-to-first-event endpoint consisting of k components can be expressed in a multistate model with k competing events. At the time origin $t = 0$, every patient is in state $Z(0) = 0$ and stays there until the occurrence of any first event. Occurrence of an event of type l is modeled by transitions into state $Z(T) = l$, $l = 1,\ldots,k$, where l is also called cause of failure and $T := \inf\{t > 0 \mid Z(t) \neq 0\}$. Thereby, it is assumed that the event times are left-truncated and right-censored. This approach only considers the time-to-*first*-event but does not model subsequent events for the reasons discussed in Sect. 2.4.

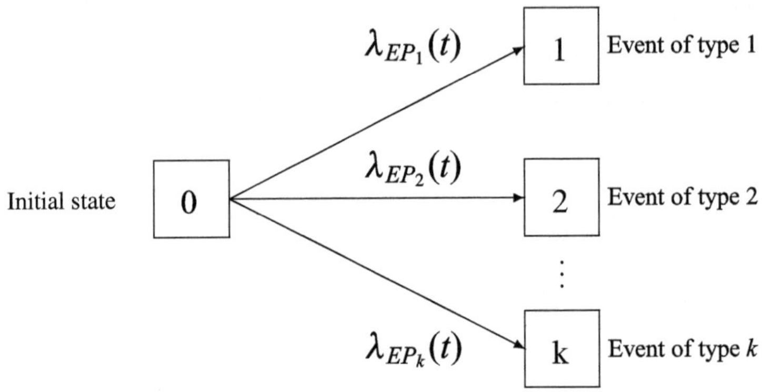

Fig. 5.1 Competing risk multistate model with corresponding cause-specific hazards

A competing risk model is usually parametrized using the cause-specific hazards (Andersen et al. 1993) given as

$$\lambda_{EP_l}(t) := \lim_{\Delta t \to 0} \frac{P\,(T \in [\,t, t + \Delta t\,)\,, Z(T) = l \,|\, T \geq t)}{\Delta t}, \; l = 1, \ldots k,$$

which are the transition hazards from state $Z(0) = 0$ into states $Z(T) = l$, $l = 1, \ldots, k$. Therefore, the cause-specific hazards define the hazards for the individual components. Note that in the definition of the cause-specific hazards, only the first occurring event is counted. This means that if another event different from type l is observed at first but an event of type l occurs subsequently, then the patient is treated as being censored at the time the first event occurs. The competing risk multistate model is illustrated in Fig. 5.1. The hazard of the composite endpoint corresponds to the so-called all-cause hazard which is defined as the sum of the cause-specific hazards

$$\lambda_{CE}(t) := \sum_{l=1}^{k} \lambda_{EP_l}(t).$$

Given an event in the composite endpoint at time t, the conditional probability that the event is of type l corresponds to $\lambda_{EP_l}(t)/\lambda_{CE}(t)$; compare Beyersmann et al. (2012).

5.2.1 *Test Problem*

The test hypotheses for the confirmatory analysis of the composite time-to-first-event endpoint are based on the all-cause hazards, where a smaller hazard corresponds to a more favorable result

$$H_0^{CE}: \frac{\lambda_{CE}^C(t)}{\lambda_{CE}^I(t)} \leq 1 \quad \text{versus} \quad H_1^{CE}: \frac{\lambda_{CE}^C(t)}{\lambda_{CE}^I(t)} > 1. \tag{5.17}$$

The test hypotheses for an individual component are formulated in terms of the cause-specific hazards

$$H_0^{EP_l,\delta_l}: \frac{\lambda_{EP_l}^C(t)}{\lambda_{EP_l}^I(t)} \leq \exp(\delta_l) \text{ versus } H_1^{EP_l,\delta_l}: \frac{\lambda_{EP_l}^C(t)}{\lambda_{EP_l}^I(t)} > \exp(\delta_l),$$
$$l = 1, \ldots, k, \ \delta_l \leq 0. \tag{5.18}$$

Thereby, as before, for $\delta_l = 0$ the test problem corresponds to a superiority test, whereas for $\delta_l < 0$ the component is tested for non-inferiority. Again note that if an individual component EP_l is chosen as the unique primary efficacy endpoint, then a superiority test problem is usually preferred, that means $\delta_l = 0$.

5.2.2 *Test Statistics*

A standard approach to test for group differences by means of the hazard ratio is the logrank test. Without loss of generality, the score test version of the logrank test will be considered in the following. Its non-inferiority version is based on the score test statistic of the Cox-model evaluated at margin δ_l (Chow et al. 2008; Jung et al. 2005). The score test statistics for the composite and its components are given as

$$W_{CE} := \frac{N_{CE}^C(\tau) - \int_0^\tau \frac{Y^C(t)}{Y^I(t)+Y^C(t)}\, d\left(N_{CE}^C(t) + N_{CE}^I(t)\right)}{\widehat{\sigma}_{CE}}, \tag{5.19}$$

$$W_{EP_l,\delta_l} := \frac{N_{EP_l}^C(\tau) - \int_0^\tau \frac{\exp(\delta_l)Y^C(t)}{Y^I(t)+\exp(\delta_l)Y^C(t)}\, d\left(N_{EP_l}^C(t) + N_{EP_l}^I(t)\right)}{\widehat{\sigma}_{EP_l,\delta_l}},$$
$$l = 1, \ldots, k, \tag{5.20}$$

with $\delta_l \leq 0$, $l = 1, \ldots, k$. Thereby, $N_{EP_l}^I(t), N_{EP_l}^C(t)$, $l = 1, \ldots, k$, denote the observed number of events of type l in the interval $[0;t]$ and $N_{CE}^I(t) = \sum_{l=1}^k N_{EP_l}^I(t)$, $N_{CE}^C(t) = \sum_{l=1}^k N_{EP_l}^C(t)$ denote the number of patients with *any* observed event in $[0,t]$ in the intervention and the control group, respectively. The parameter τ defines the largest uncensored event time. Moreover, the risk set at time t defined as the number of patients in state 0 just before time t is denoted by $Y^I(t)$ and $Y^C(t)$, respectively. The integral expressions correspond to the expected number of events in the control group under the corresponding null hypothesis. Under the proportional hazards assumption, the variance estimators from the Cox-model are given by

$$\widehat{\sigma}^2_{CE} = \int_0^\tau \frac{Y^I(t)Y^C(t)}{\left(Y^I(t) + Y^C(t)\right)^2}\, d\left(N^C_{CE}(t) + N^I_{CE}(t)\right),$$

$$\widehat{\sigma}^2_{EP_l,\delta_l} = \int_0^\tau \frac{\exp(\delta_l)Y^I(t)Y^C(t)}{\left(Y^I(t) + \exp(\delta_l)Y^C(t)\right)^2}\, d\left(N^C_{EP_l}(t) + N^I_{EP_l}(t)\right).$$

According to Chow et al. (2008) and Schoenfeld (1981), the test statistics (5.19) and (5.20) are approximately distributed as

$$W_{CE} \sim N\left(\ln\left(\theta_{CE}\right)\sqrt{\frac{n_I \cdot n_C}{n_I + n_C} \cdot \pi_{CE}}, 1\right), \tag{5.21}$$

$$W_{EP_l,\delta_l} \sim N\left(\ln\left(\frac{\theta_{EP_l}}{\exp\left(\delta_l\right)}\right)\sqrt{\frac{n_I \cdot n_C}{n_I + n_C} \cdot \pi_{EP_l}}, 1\right), \tag{5.22}$$

where $\theta_{CE} := \lambda^C_{CE}(t)/\lambda^I_{CE}(t), \theta_{EP_l} := \lambda^C_{EP_l}(t)/\lambda^I_{EP_l}(t)$ denote the corresponding true hazard ratios and π_{CE}, π_{EP_l}, $l = 1, \ldots, k$, denote the total probabilities for a corresponding event in the pooled groups. Therefore, the null hypotheses are rejected whenever $W_{CE} \geq z_{1-\alpha}$ or $W_{EP_l,\delta_l} \geq z_{1-\alpha}$, respectively, where α denotes the one-sided significance level.

5.2.3 *Sample Size Calculation*

As before for binary endpoints, in this section the standard approach for sample size calculation and robust alternatives will be discussed. Chapter 6 moreover presents adaptive and group-sequential designs for time-to-event endpoints to deal with planning uncertainties.

5.2.3.1 Standard Sample Size Calculation

The sample size formula for the superiority test problem formulated in (5.17) can be deduced by solving

$$P_{H_1^{CE}}\left(W_{CE} \geq z_{1-\alpha}\right) \geq 1 - \beta. \tag{5.23}$$

It can easily be seen that for a predefined effect $\theta_{CE} := \lambda_{CE}^{C}(t)/\lambda_{CE}^{I}(t)$ under H_1^{CE} solving (5.23) is equivalent to

$$1 - P_{H_1^{CE}}\left(X \leq z_{1-\alpha} - \ln(\theta_{CE})\sqrt{\frac{n_I \cdot n_C}{n_I + n_C}}\right) \geq 1 - \beta, \text{ for } X \sim N(0,1)$$

$$\Leftrightarrow z_{1-\alpha} - \ln(\theta_{CE})\sqrt{n_I \cdot \left(\frac{1}{\gamma + 1}\right) \cdot \pi_{CE}} \leq -z_{1-\beta},$$

$$\Leftrightarrow n_I \geq (\gamma + 1) \cdot \frac{\left(z_{1-\alpha} + z_{1-\beta}\right)^2}{\ln(\theta_{CE})^2 \cdot \pi_{CE}},$$

where $\gamma = n_I/n_C$ is the sample size allocation ratio. The total sample size for both groups is given by the smallest integer $n_I + n_C$ fulfilling

$$n_I + n_C \geq \left(\gamma + \frac{1}{\gamma} + 2\right) \cdot \frac{\left(z_{1-\alpha} + z_{1-\beta}\right)^2}{\ln(\theta_{CE})^2 \cdot \pi_{CE}}. \tag{5.24}$$

It can be seen that the required sample size depends on the overall probability π_{CE} for an event in the composite in the pooled sample of both groups which is given as

$$\pi_{CE} = \frac{n_I}{n_I + n_C} \cdot \pi_{CE}^{I} + \frac{n_C}{n_I + n_C} \cdot \pi_{CE}^{C}, \tag{5.25}$$

where $\pi_{CE}^{I}, \pi_{CE}^{C}$ are the probabilities for an event in each of the groups. Sample size calculation for a time-to-event outcome is performed in two steps:

1. Calculate the required total number of events in both groups $d_{CE} := (n_I + n_C) \cdot \pi_{CE}$ which is given as the smallest integer fulfilling

$$d_{CE} \geq \left(\gamma + \frac{1}{\gamma} + 2\right) \cdot \frac{\left(z_{1-\alpha} + z_{1-\beta}\right)^2}{\ln(\theta_{CE})^2}. \tag{5.26}$$

2. Calculate the corresponding total sample size providing the required number of events as

$$n_I + n_C := \frac{d_{CE}}{\pi_{CE}}. \tag{5.27}$$

For step 2, a reasonable estimate of π_{CE} is required which depends on the accrual duration, on the accrual distribution, on the minimal follow-up, on the underlying time-to-event distribution, and on the corresponding hazards as well as on the censoring distribution. In a standard clinical trial setting, it is usually assumed that patient accrual is uniformly distributed over a predefined accrual interval $[0; a]$, where a corresponds to the accrual duration. The minimal follow-up per patient

f is also fixed in advance. Assuming exponentially distributed event times in each group, the probabilities for an event in each group are given as (Schulgen et al. 2005)

$$\pi_{CE}^{I} = 1 - \frac{\left(\exp\left(-\lambda_{CE}^{I} \cdot f\right) - \exp\left(-\lambda_{CE}^{I} \cdot (a+f)\right)\right)}{\lambda_{CE}^{I} \cdot a}, \tag{5.28}$$

$$\pi_{CE}^{C} = 1 - \frac{\left(\exp\left(-\lambda_{CE}^{C} \cdot f\right) - \exp\left(-\lambda_{CE}^{C} \cdot (a+f)\right)\right)}{\lambda_{CE}^{C} \cdot a}. \tag{5.29}$$

For arbitrary event time distributions $S_I(t), S_C(t)$, the probabilities might be approximated using Simpson's rule (Schoenfeld 1983)

$$\pi_{CE}^{I} \approx 1 - \frac{1}{6} \cdot \left(S_I(f) + 4 \cdot S_I(f + 0.5 \cdot a) + S_I(f+a)\right), \tag{5.30}$$

$$\pi_{CE}^{C} \approx 1 - \frac{1}{6} \cdot \left(S_C(f) + 4 \cdot S_C(f + 0.5 \cdot a) + S_C(f+a)\right). \tag{5.31}$$

The above approach takes account of administrative censoring due to a limited observational period. An ad hoc method to take account of additionally censored observations is to divide the final total sample size by an assumed fraction of censored observations at the final analysis time point. When a specific censoring distribution should be incorporated, the probability for a censored observation at an arbitrary time point can be estimated using the above formulas (5.28)–(5.31). This is especially important, if competing types of events are likely to influence the expected proportion of the event under consideration. General aspects of competing risks are discussed in Part IV in this book.

The sample size formula for the (non-inferiority) test problem formulated in (5.18) for an individual component EP_l, $l = 1, \ldots, k$, can be deduced equivalently (compare Chow et al. (2008)) and is given by

$$n_I + n_C \geq \left(\gamma + \frac{1}{\gamma} + 2\right) \cdot \frac{\left(z_{1-\alpha} + z_{1-\beta}\right)^2}{\left(\ln\left(\theta_{EP_l}\right) - \delta_l\right)^2 \cdot \pi_{EP_l}}. \tag{5.32}$$

The calculation of π_{EP_l} is equivalent as for π_{CE}.

5.2.3.2 Robust Sample Size Calculation

Again, an easy approach to get a conservative estimate of the required sample size would be to calculate the classical frequentist power for a range of different plausible parameter settings and to base the sample size calculation on the worst-case scenario which is based on the highest sample size. However, as discussed before for binary endpoints, this might yield unnecessarily high sample sizes. The expected power approach proposed for composite binary endpoints is more complicated in a time-to-event setting, as the distribution of the power function depends not only on the underlying hazards but also on the expected proportion of events, which is again a function of the corresponding hazards and, if applicable, on the competing hazards of other event types. Integration of the power with respect to given priors for the underlying hazards therefore corresponds to a higher computational effort. For this reason, in the case of planning uncertainties in a time-to-event setting, we recommend the use of group-sequential or adaptive designs which are further discussed in Chap. 6.

5.3 Time-to-Event Endpoints for Non-proportional Hazards

Non-proportional hazards are commonly observed in clinical applications. The class of event time distributions fulfilling the proportional hazards assumption is generally limited, and there are many other distributions for which the hazards are non-proportional. Moreover, there exist situations where the proportional hazards assumption cannot hold true by construction of the underlying test problem. When a composite time-to-first-event endpoint and its components are analyzed, the proportional hazards assumption cannot simultaneously hold true for all endpoints, except for the special case of equal baseline hazards which is in particular fulfilled for constant hazards (exponential distributions). Therefore, methods for non-proportional hazards are of particular interest within the context of composite endpoints.

Under non-proportional hazards, the test problem for a confirmatory analysis of a composite time-to-first-event endpoint can be formulated in terms of the so-called average hazard ratio which is defined as

$$\Psi_{CE} := \frac{\int_0^\infty \frac{\lambda_{CE}^C(t)}{\lambda_{CE}^C(t)+\lambda_{CE}^I(t)}\, dG(t)}{\int_0^\infty \frac{\lambda_{CE}^I(t)}{\lambda_{CE}^C(t)+\lambda_{CE}^I(t)}\, dG(t)}, \tag{5.33}$$

where $\lambda_{CE}^I(t)$, $\lambda_{CE}^C(t)$ denote the all-cause hazards (compare Sect. 5.2) and $G(t)$ is a given event time distribution corresponding to a weighting function. Equivalently, the average hazard ratio for an individual component is given as

$$\Psi_{EP_l} := \frac{\int_0^\infty \frac{\lambda_{EP_l}^C(t)}{\lambda_{EP_l}^C(t)+\lambda_{EP_l}^I(t)}\, dG(t)}{\int_0^\infty \frac{\lambda_{EP_l}^I(t)}{\lambda_{EP_l}^C(t)+\lambda_{EP_l}^I(t)}\, dG(t)}, \quad l = 1, \ldots, k, \tag{5.34}$$

where $\lambda_{EP_l}^I(t)$, $\lambda_{EP_l}^C(t)$ denote the cause-specific hazards as defined in Sect. 5.2. The average hazard ratio as defined above is independent of time. However, the observation time L within a clinical trial is always limited, and therefore the average hazard ratio can only be consistently estimated within the interval $[0; L]$. Therefore, a truncated version of the average hazard ratio can be defined as

$$\Psi_{CE}(L) := \frac{\int_0^L \frac{\lambda_{CE}^C(t)}{\lambda_{CE}^C(t)+\lambda_{CE}^I(t)}\, dG(t)}{\int_0^L \frac{\lambda_{CE}^I(t)}{\lambda_{CE}^C(t)+\lambda_{CE}^I(t)}\, dG(t)}, \tag{5.35}$$

$$\Psi_{EP_l}(L) := \frac{\int_0^L \frac{\lambda_{EP_l}^C(t)}{\lambda_{EP_l}^C(t)+\lambda_{EP_l}^I(t)}\, dG(t)}{\int_0^L \frac{\lambda_{EP_l}^I(t)}{\lambda_{EP_l}^C(t)+\lambda_{EP_l}^I(t)}\, dG(t)}, \quad l = 1, \ldots, k. \tag{5.36}$$

The truncated version of the average hazard ratio thus depends on the observation time. This, however, cannot be considered as a disadvantage of the average hazard ratio but is a general property of any effect measure under non-proportional hazards when the observation time is limited (Royston and Parmar 2011). In particular, the common hazard ratio under non-proportional hazards is also a function of time. In applications, L should be chosen reasonably large with respect to the expected event time distributions. For sufficiently large L, $\Psi_{CE}(L)$ and $\Psi_{EP_i}(L)$, $i = 1, \ldots, k$, will approach Ψ_{CE} and Ψ_{EP_l}, $l = 1, \ldots, k$, respectively. For the sake of simplicity, we only consider the truncated version of the average hazard ratio thereby assuming that L is sufficiently large.

Note that in the original work of Kalbfleisch and Prentice (1981), the average hazard ratio is not introduced as a relative effect measures between groups, but the nominator of (5.35) or (5.36), respectively, multiplied by a factor $-(1 - G(L))^{-1}$ is used as an absolute effect measure. By definition, for $L \leq \infty$, the nominator and the denominator of (5.35) sum up to $G(0) - G(L) \leq 1$, and therefore both approaches can be used equivalently. Without loss of generality, we will use the definition of the average hazard ratio as a relative effect measure in order to be more directly comparable to the standard hazard ratio.

In the original work of Kalbfleisch and Prentice (1981), the following class of weighting functions was introduced

$$G(t) := S_I^\nu(t) S_C^\nu(t), \ \nu > 0, \tag{5.37}$$

where $S_I(t), S_C(t)$ are the event time distributions for the two groups and ν corresponds to a shape parameter which determines the magnitude of the weights over time. Rauch et al. (2017) have shown that the weights proposed by Kalbfleisch and Prentice (1981) with $\nu = 1$ have favorable properties for a range of possible underlying event time distributions and can therefore quite generally be recommended. As an alternative approach, Schemper et al. (2009) defined another weighting function which, however, yields comparable results and has similar performance properties as the latter approach. Other choices of ν for the weights proposed by Kalbfleisch and Prentice (1981) should only be used with care. Depending on the underlying event time distributions, $\nu \leq 0.5$ often results in a high standard deviation of the corresponding effect estimator (introduced below) and might moreover also cause systematic bias. Similarly, setting $\nu \geq 1$ usually also yields a high standard deviation of the effect estimator. Therefore, in the remainder of this book, the average hazard ratio is always defined by means of the following weights

$$G(t) := S_I(t)S_C(t). \tag{5.38}$$

The average hazard ratios (5.35) and (5.36) then simplify to

$$\Psi_{CE}(L) := \frac{\int_0^L S_{CE}^I(t)\, d\left(S_{CE}^C(t)\right)}{\int_0^L S_{CE}^C(t)\, d\left(S_{CE}^I(t)\right)}, \tag{5.39}$$

$$\Psi_{EP_l}(L) := \frac{\int_0^L S_{EP_l}^I(t)\, d\left(S_{EP_l}^C(t)\right)}{\int_0^L S_{EP_l}^C(t)\, d\left(S_{EP_l}^I(t)\right)}, \ l = 1, \ldots, k. \tag{5.40}$$

Setting

$$\Xi_{CE}(L) := \ln(\Psi_{CE}(L)), \tag{5.41}$$

$$\Xi_{EP_l}(L) := \ln(\Psi_{EP_l}(L)), \ l = 1, \ldots, k, \tag{5.42}$$

intuitive estimators $\widehat{\Xi}_{CE}(L), \widehat{\Xi}_{EP_l}(L), \ l = 1, \ldots, k$, are obtained by substituting the underlying event time distributions by their Kaplan-Meier estimates. Corresponding variance estimators $\widehat{\sigma}^2_{\Xi_{CE}(L)}, \widehat{\sigma}^2_{\Xi_{EP_l}(L)}$ were implemented in the R package AHR by Brückner (2016).

5.3.1 *Test Problem*

The test hypotheses in terms of the average hazard ratio for the composite endpoint are given as

$$H_0^{CE} : \Psi_{CE}(L) \leq 1 \quad \text{versus} \quad H_1^{CE} : \Psi_{CE}(L) > 1. \tag{5.43}$$

For the individual components, the test hypotheses read as

$$H_0^{EP_l} : \Psi_{EP_l}(L) \leq 1 \quad \text{versus} \quad H_1^{EP_l} : \Psi_{EP_l}(L) > 1, \ l = 1, \ldots, k. \tag{5.44}$$

5.3.2 Test Statistics

To test the above hypotheses, Wald-type test statistics can be defined as

$$Y_{CE} := -\frac{\widehat{\Xi}_{CE}(L)}{\sqrt{\widehat{\delta}^2_{\Xi_{CE}(L)}}}, \tag{5.45}$$

$$Y_{EP_l} := -\frac{\widehat{\Xi}_{EP_l}(L)}{\sqrt{\widehat{\delta}^2_{\Xi_{EP_l}(L)}}}, \ l = 1, \ldots, k, \tag{5.46}$$

These test statistics are approximately distributed as

$$Y_{CE} \sim N\left(-\ln(\Psi_{CE}(L)), 1\right), \tag{5.47}$$

$$Y_{EP_l} \sim N\left(-\ln(\Psi_{EP_l}(L)), 1\right), \ l = 1, \ldots, k. \tag{5.48}$$

Therefore, the null hypothesis is rejected whenever $Y_{CE} \geq z_{1-\alpha}$ respective $Y_{EP_l} \geq z_{1-\alpha}$, $l = 1, \ldots, k$. The above test will be denoted as the average hazard ratio test in the remainder of this book. Note that the negative signs in the test statistics (5.45) and (5.46) were added, without loss of generality, to achieve consistency for the critical values referring to the different test statistics introduced within this book.

5.3.3 Sample Size Calculation

Sample size calculation for the average hazard ratio test is not straightforward as the magnitude of the average hazard ratio and its variance depend on the true underlying event time distributions which are usually not known in the planning stage. For given event time distributions under the alternative hypothesis and a fixed follow-up time L, the power for a given sample size can be simulated. An R code to approximate the required sample size to reach a predefined power for Weibull-distributed event time distributions is provided in Chap. 7. Note that due to the sensible dependence of the sample size on the correct specification of the underlying event time distributions, we recommend to consider a range of plausible scenarios and take the corresponding maximal sample size in order to gain some robustness.

5.4 Recurrent Event Analysis

Composite endpoints are usually analyzed by taking only the first occurring event into account. Thereby, it is ignored that an individual may have more than one nonfatal event of interest which clearly defines an important loss of information. Therefore, methods for recurrent event analysis are generally of interest within the context of composite endpoints. A simple approach to analyze several events per patient is to count every event of each individual within the study period and model these count data, e.g., by means of Poisson or negative binomial regression (Wang et al. 2009). This simple formulation of a recurrent event process does not account for event times, loss to follow-up (censoring), or competing risks. To model recurrent event times, there exist several approaches based on the common Cox-model such as the model by Andersen and Gill (1982), the marginal model from Wei et al. (1989), or conditional models as proposed by Prentice et al. (1981).

In the Andersen-Gill model, every individual belongs to the risk set for a subsequent event, irrespective of the fact whether a first event has already occurred unless the patient time was not censored. This implies that all event times are treated as independent irrespective whether the events belong to the same or to different patients. In other words, the hazard to experience a first event at time t is assumed to be the same as the hazard to experience a second (or any subsequent) event at time t. This is a very restrictive assumption. In clinical practice, the risk for a subsequent event usually changes after occurrence of the first. For example, the risk to die increases once a myocardial infarction occurred.

The model by Wei et al. (1989) is a stratified model, where the first stratum contains all first event times, strata two all second event times, and so on. Like for the Andersen-Gill model, every individual belongs to the risk set for a subsequent event, irrespective of the fact whether a first event has already occurred. This model allows for different baseline hazards between the strata, but the hazard ratios within each strata are assumed to be the same. By this, the model is less restrictive than the Andersen-Gill model. However, the strata for the subsequent events are often small, and as a consequence the estimated stratified effects are based on a high level of uncertainty.

Like the model by Wei et al. (1989), the models of Prentice et al. (1981) are based on stratification. The two models by Prentice et al. (1981) differ with respect to the underlying time scale. The first model considers the time since study entry, whereas the second uses the time since the previous event which is referred to as the "gap time model" in the literature (Cook and Lawless 2007). Both models take into account that a subsequent event can only occur after experiencing a previous event, thus not every individual is directly under risk for any subsequent event like in the other approaches presented above.

Other general extensions of Cox-model based approaches are given by frailty models which include a random factor to account for unobserved covariates or heterogeneity between individuals (Cook and Lawless 2007). Thereby, several event times belonging to the same individual are assumed to be more similar than event times belonging to different patients. However, the hazard or the hazard ratio, respectively, for the first event are still assumed to be equivalent to the hazard or hazard ratio of any subsequent event.

As a further alternative, a recurrent event process can also be modeled through a corresponding multistate model by estimating each transition hazard separately thereby allowing different hazards for each transition (Cook and Lawless 2007). However, a multistate model does not yield a single summary effect measure. Moreover, the complexity of these models is often quite high, and to estimate all transition hazards within a multistate model with reasonable accuracy, the required sample size soon becomes unrealistically large.

None of the above approaches allows to model the case that the hazard for a subsequent event depends on the observational time point of the previous event. In clinical practice, however, it is reasonable that the risk for further events is different for patients who experience no event for a long time as compared to patients that directly have an event. For example, the risk for a second myocardial infarction after a first one may be higher for patient who soon experience a first infarction than for patient who are event-free for a long time.

To summarize, modeling recurrent events is always based on quite restrictive assumptions which are usually not met in practice. If the required assumptions are not fulfilled, the estimated treatment effect is difficult to predict and thus even more difficult to interpret. Therefore, the current practice to concentrate on the analysis of the first occurring event seems justified. As discussed within Chap. 2 in Part I of this book, there already exist a number of interpretation problem concerning a standard composite effect when considering only the first event. Dealing with all possible consequences of misspecified recurrent event models goes beyond the scope of this book. However, we do recommend to consider recurrent event models whenever the underlying assumptions seem justified.

References

Andersen, P., Borgan, Ø., Gill, R., & Keiding, N. (1993). *Statistical models based on counting processes*. New York: Springer.

Andersen, P. K., & Gill, R. D. (1982). Cox's regression model for counting processes: A large sample study. *The Annals of Statistics, 10*, 1100–1120.

Begum, N., & King, M. L. (2005). Most mean powerful test of a composite null against a composite alternative. *Computational Statistics & Data Analysis, 49*, 1079–1104.

Beyersmann, J., Allignol, A., & Schumacher, M. (2012). *Competing risks and multistate models with R*. New York: Springer.

Blackwelder, W. C. (1982). Proving the null hypothesis in clinical trials. *Controlled Clinical Trials, 3*, 345–353.

Brown, B. W., Herson, J., Atkinson, N. E., & Rozell, E. M. (1987). Projection from previous studies: A Bayesian and frequentist compromise. *Controlled Clinical Trials, 8*, 29–44.

Brückner, M. (2016). *AHR: Estimation and testing of average hazard ratios*. R package version 1.4.2. https://CRAN.R-project.org/package=AHR. Accessed April 9, 2018.

Chow, S., Shao, J., & Wang, H. (2008). *Sample size calculations in clinical research*. Boca Raton: Chapman & Hall.

Cook, R. J., & Lawless, J. F. (2007). *The statistical analysis of recurrent events*. New York: Springer.

Daimon, T. (2008). Bayesian sample size calculations for a non-inferiority test of two proportions in clinical trials. *Contemporary Clinical Trials, 29*, 507–516.

Dunnett, C. W., & Gent, M. (1977). Significance testing to establish equivalence between treatments, with special reference to data in the form of 2x2 tables. *Biometrics, 33*, 593–602.

Farrington, C. P., & Manning, G. (1990). Test statistics and sample size formulae for comparative binomial trials with null hypothesis of non-zero risk difference or non-unity relative risk. *Statistics in Medicine, 9*, 1447–1454.

Gillett, R. (1994). An average power criterion for sample size estimation. *The Statistician, 43*, 389–394.

Jung, S., Kang, S., McCall, L., & Blumenstein, B. (2005). Sample size computation for two-sample noninferiority log-rank test. *Journal of Biopharmaceutical Statistics, 15*, 969–979.

Kalbfleisch, J. D., & Prentice, R. L. (1981). Estimation of the average hazard ratio. *Biometrika, 68*, 105–112.

Prentice, R. L., Williams, B. J., & Peterson, A. V. (1981). On the regression analysis of multivariate failure time data. *Biometrika, 68*, 373–379.

Rauch, G., Brannath, W., Brückner, M., & Kieser, M. (2017). The average hazard ratio - A good effect measure for time-to-event endpoints when the proportional hazard assumption is violated? *Methods of Information in Medicine* (accepted).

Rauch, G., & Kieser, M. (2013). An expected power approach for the assessment of composite endpoints and their components. *Computational Statistics & Data Analysis, 60*, 111–122.

Roebruck, P., & Kühn, A. (1995). Comparison of tests and sample size formulae for proving therapeutic equivalence based on the difference of binomial probabilities. *Statistics in Medicine, 14*, 1583–1594.

Royston, P., & Parmar, M. K. B. (2011). The use of restricted mean survival time to estimate the treatment effect in randomized clinical trials when the proportional hazards assumption is in doubt. *Statistics in Medicine, 30*, 2409–2421.

Schemper, M., Wakounig, S., & Heinze, G. (2009). The estimation of average hazard ratios by weighted Cox regression. *Statistics in Medicine, 28*, 2473–2489.

Schoenfeld, D. (1981). The asymptotic properties of nonparametric tests for comparing survival distributions. *Biometrika, 68*, 316–319.

Schoenfeld, D. (1983). Sample-size formula for the proportional-hazards regression model. *Biometrics, 39*, 499–503.

Schulgen, G., Olschewski, M., Krane, V., Wanner, C., Ruf, G. & Schumacher, M. (2005). Sample sizes for clinical trials with time-to-event endpoints and competing risks. *Contemporary Clinical trials, 26*, 386–396.

Shao, Y., Mukhi, V., & Goldberg, J. D. (2008). A hybrid Bayesian-frequentist approach to evaluate clinical trial designs for tests of superiority and non-inferiority. *Statistics in Medicine, 27*, 504–519.

Spiegelhalter, D. J., & Keith, R. A. (2004). *Bayesian approaches to clinical trials and health-care evaluation*. Chichester: Wiley.

Wang Y. C., Meyerson, L., Tang, Y. Q., & Qian, N. (2009). Statistical methods for the analysis of relapse data in MS clinical trials. *Journal of the Neurological Sciences, 285*, 206–211.

Wei, L. J., Lin, D. Y., & Weissfeld, L. (1989). Regression analysis of multivariate incomplete failure time data by modeling marginal distributions. *Journal of the American Statistical Association, 84*, 1065–1073.

Chapter 6
Group-Sequential and Adaptive Designs

As motivated in Part I of this book, a major difficulty when planning a clinical trial with a composite primary endpoint or, as an alternative, with a single component as primary efficacy measure is to make valid planning assumptions on the required parameters and the expected effect size. A common way to deal with planning uncertainties is the use of adaptive or classical group-sequential designs, which allow an early stop of the trial either for futility or, in case of an unexpectedly large effect, also for efficacy. It is intuitively clear that the option of an early termination of the trial can decrease the total required sample size by a relevant amount. In adaptive designs, it is moreover possible to change other trial settings such as the second-stage sample size or the test strategy. As repeated hypothesis tests are performed in group-sequential and adaptive designs, the local significance levels for each individual analysis have to be adjusted in order to maintain the global type I error rate. Therefore, under the correct parameter assumptions, the maximal required sample size of such a design is larger than the sample size for a standard single-stage design.

In this chapter, (adaptive) group-sequential designs for a single primary endpoint will be introduced. Throughout this book, the expression "group-sequential design" will be used for designs incorporating preplanned interim analyses but without the option to modify design parameters at interim, whereas the expression "adaptive group-sequential design" or simply "adaptive design" will be used, whenever adaptive modifications might be incorporated. Note that these expressions are not always used consistently in the corresponding literature.

As before, the underlying test problems will be formulated separately for (composite) binary endpoints and (composite) time-to-(first-)event endpoints. As the methodology is equivalent for a composite and a single primary endpoint, it is sufficient to introduce the test problem for the composite. For the sake of simplicity, a two-stage approach will be considered with one interim analysis after recruitment of a total of n_1 patients. The final analysis will take place when $n = n_1 + n_2$ patients have been recruited and are fully observed. Moreover, in this section equal group sizes per stage will be assumed.

G. Rauch et al., *Planning and Analyzing Clinical Trials with Composite Endpoints*, Springer Series in Pharmaceutical Statistics,
https://doi.org/10.1007/978-3-319-73770-6_6

6.1 Stage-Wise Local Levels

The general idea of two-stage group-sequential and adaptive designs is to define adequate test statistics Z^1 and Z^{1+2} for the interim analysis and for the final analysis, respectively, for which the multivariate joint distribution can be determined in the planning stage. For all applications in this book, this joint distribution is given by a bivariate normal distribution which also corresponds to the standard case considered in the literature. The test statistic Z^{1+2} includes all data collected until the final analysis and is thus positively correlated to Z^1. At the interim analysis, after recruitment of n_1 patients, the null hypothesis is rejected whenever

$$Z^1 \geq z_{1-\alpha_1}, \tag{6.1}$$

where $z_{1-\alpha_1}$ denotes the corresponding quantile of the standard normal distribution and α_1 is the adjusted local significance level of the first stage. Otherwise, the trial is continued, and an additional number of n_2 patients are recruited. Thereby, in a nonadaptive group-sequential design, n_2 is fixed in the planning stage, whereas in an adaptive design with sample size recalculation, n_2 corresponds to a random variable. At the final analysis, the null hypothesis is rejected if

$$Z^{1+2} \geq z_{1-\alpha_{1+2}}. \tag{6.2}$$

The local levels α_1 and α_{1+2} are thereby defined such that the overall type I error is controlled by the global significance level α, that is

$$P_{H_0^{CE}}\left(Z^1 \geq z_{1-\alpha_1} \vee (Z^1 < z_{1-\alpha_1} \wedge Z^{1+2} \geq z_{1-\alpha_{1+2}})\right) = \alpha. \tag{6.3}$$

Due to the underlying multiple test problem when using a group-sequential or adaptive design, the maximal possible sample size n is always larger than the required sample size for a corresponding one-stage design. However, as the finally required sample size of a group-sequential or adaptive design is a random variable, considering only the maximal sample size is not a fair performance comparison to a one-stage design where the sample size is fixed. Instead, the average sample size under the corresponding null hypothesis or under a specified alternative is more appropriate to assess the design properties.

6.1.1 Choice of Stage-Wise Local Levels

Due to the positive correlation of the test statistics Z^1 and Z^{1+2}, the sum of the local levels may be larger than the global significance level while still controlling the global type I error rate. The local significance levels in a group-sequential design can be chosen in various ways. Pocock (1977) proposed to choose the same local levels for each analysis, i.e., $\alpha_1 = \alpha_{1+2}$. Another strategy would be to choose lower local levels at earlier time points and higher local levels at later stages motivated by the fact that at later stages the number of recruited patients is higher and rejection of the null hypothesis is thus more likely. Increasing local levels $\alpha_1 < \alpha_{1+2}$ have been proposed by various authors. A popular approach was published by O'Brien and Fleming (1979). However, there are many other possibilities to define appropriate local levels by the use of predefined alpha-spending functions; compare, for example, Hwang et al. (1990), Kim and Demets (1987), and Lan and DeMets (1983). In the context of a composite endpoint with high planning uncertainties, it is desirable to provide a reasonable chance to stop the study early with proof of efficacy in case the treatment effect is larger than originally anticipated. This could motivate to already apply rather higher local levels at early stages (e.g., Pocock boundaries) instead of increasing local levels over time. However, the specific clinical trial scenario must be considered to define an adequate choice of a meaningful alpha-spending function. An overview on group-sequential designs is given by Wassmer (1999) and Jennison and Turnbull (1999).

In the following, the corresponding stage-wise test statistics for binary and time-to-event endpoints are defined.

6.2 Stage-Wise Test Statistics for Binary Endpoints

Let T_{CE}^1 denote the test statistic for the composite endpoint after recruitment of n_1 patients, that is, after stage 1 of the trial, and let T_{CE}^{1+2} denote the test statistic at the final analysis after recruitment of the total number of $n = n_1 + n_2$ patients. The test statistic T_{CE}^{1+2} includes all data of stage 1 and 2 and is therefore positively correlated to T_{CE}^1 . Under the corresponding null hypothesis, these test statistics T_{CE}^1 and T_{CE}^{1+2} are standard normally distributed. The test statistic T_{CE}^{1+2} including all data collected until the final analysis can be written as

$$T_{CE}^{1+2} = \frac{\sqrt{n_1} \cdot T_{CE}^1 + \sqrt{n_2} \cdot T_{CE}^2}{\sqrt{n_1 + n_2}}, \tag{6.4}$$

where T_{CE}^2 is the independent incremental test statistic for stage 2 constituted exclusively of the data collected in the second stage. Therefore, the covariance between T_{CE}^1 and T_{CE}^{1+2} is given by

$$Cov\left(T_{CE}^1, T_{CE}^{1+2}\right) = \sqrt{\frac{n_1}{n_1 + n_2}} = \sqrt{\psi}, \tag{6.5}$$

where ψ corresponds to the information fraction at interim. By using this joint distribution given as a bivariate normal distribution with variance-covariance matrix according to (6.5), adjusted local levels α_1 and α_{1+2} which guarantee that the overall type I error is controlled by the global significance level α can be defined by solving

$$P_{H_0^{CE}}\left(T_{CE}^1 \geq z_{1-\alpha_1} \vee (T_{CE}^1 < z_{1-\alpha_1} \wedge T_{CE}^{1+2} \geq z_{1-\alpha_{1+2}})\right) = \alpha. \tag{6.6}$$

In a standard group-sequential design, the sample size for the second-stage n_2 is fixed in the planning stage. In contrast, adaptive group-sequential designs allow flexible modifications of trial parameters, such as the second-stage sample size, at the interim stage. For this purpose, the test statistics for stage 1 and 2 have to be defined in a similar but more general way

$$\widetilde{T}_{CE}^1 := T_{CE}^1,\ \widetilde{T}_{CE}^{1+2} := \frac{w_1 \cdot T_{CE}^1 + w_2 \cdot \widetilde{T}_{CE}^2}{\sqrt{w_1^2 + w_2^2}}, \tag{6.7}$$

where again $\widetilde{T}_{CE}^2$ is the independent incremental test statistic for stage 2 constituted exclusively of the data collected in the second stage, which is now allowed to be based on design modifications, e.g. on a different stage 2 sample size. The weights $w_1, w_2 \in [0; \infty)$ are predefined in the planning stage. The covariance between these alternative test statistics now exclusively depends on the predefined weights w_1, w_2 and is given by

$$Cov\left(\widetilde{T}_{CE}^1, \widetilde{T}_{CE}^{1+2}\right) = \frac{w_1}{\sqrt{w_1^2 + w_2^2}}. \tag{6.8}$$

Thus, standard group-sequential boundaries can be applied to the alternative test statistics $\widetilde{T}_{CE}^1, \widetilde{T}_{CE}^{1+2}$, but modifications of the second-stage sample size at interim are now allowed. For example, choosing $w_1 = \sqrt{n_1},\ w_2 = \sqrt{n_2}$ would correspond to a weighting of the two stages according to the standard group-sequential approach which is optimal in case the sample sizes for both stages are not adapted an thus given by n_1 and n_2. In the latter case, the adaptive and the group-sequential approaches are equivalent. If the sample size n_2 is modified at interim, then the second-stage data are down-weighted if the new sample size $n_{2,recalc}$ is larger than n_2 and up-weighted for $n_{2,recalc} < n_2$.

In case a single binary component is chosen as primary endpoint, the definition of a corresponding two-stage design is completely equivalent and will therefore be omitted.

6.3 Stage-Wise Test Statistics for Time-to-Event Endpoints

When using a two-stage design for a time-to-event endpoint, an important aspect is that the observational period is usually much longer than for a binary endpoint. Therefore, in most applications it will not be feasible to stop the recruitment after inclusion of n_1 patient and to wait until all patients have been fully observed, that is, until either an event occurs or until the minimal follow-up duration is reached.

The independent incremental approach presented above can be used to take account of this issue. In the following, the stage-wise test statistics will be introduced for the common logrank test statistic which are optimal under proportional hazards as introduced in Sect. 5.2. However, the test statistic of the average hazard ratio test introduced in Sect. 5.3 also fulfills the independent incremental property (Brückner and Brannath 2016). Thus, the approach described below can be applied equivalently to the average hazard ratio test statistics.

Let W_{CE}^1 denote the test statistic for the composite endpoint after recruitment of n_1 patients, that is, after stage 1 of the trial, and let W_{CE}^{1+2} denote the test statistic at the final analysis after recruitment of the total number of $n = n_1 + n_2$ patients. Under the corresponding null hypothesis, these test statistics W_{CE}^1, W_{CE}^{1+2} are standard normally distributed. The test statistic W_{CE}^{1+2} including all data collected until the final analysis can be written as

$$W_{CE}^{1+2} := \frac{\sqrt{d_1} \cdot W_{CE}^1 + \sqrt{d_2} \cdot W_{CE}^2}{\sqrt{d_{1+2}}}, \tag{6.9}$$

where d_1, d_2, and d_{1+2} denote the total number of events observed at the interim, within the second stage and at the final analysis, respectively. The expression W_{CE}^2 is the independent incremental test statistic for stage 2 constituted exclusively of the data collected in the second stage. By reversing (6.9) it follows

$$W_{CE}^2 = \frac{\sqrt{d_{1+2}} \cdot W_{CE}^{1+2} - \sqrt{d_1} \cdot W_{CE}^1}{\sqrt{d_{1+2} - d_1}}. \tag{6.10}$$

It should be noted that W_{CE}^2 might include patients recruited during the first stage, which have not been fully observed at the interim stage.

The covariance between the first- and the second-stage test statistics is given by

$$Cov\left(W_{CE}^1, W_{CE}^{1+2}\right) = \sqrt{\frac{d_1}{d_{1+2}}} = \sqrt{\psi}, \tag{6.11}$$

where ψ is the observed information fraction at interim. As above, by using this joint distribution, adjusted local levels α_1 and α_{1+2} can be defined which guarantee that the overall type I error is controlled by the global significance level α.

Note that in order to apply a standard group-sequential design for a time-to-event endpoint as defined above, the observed proportion of events at interim must be correctly specified in the planning stage, which seems rather unrealistic. Of course, the observed proportion of events at interim can be replaced by the corresponding expected proportion $\widehat{\psi}$ estimated in the planning stage. If, however, the expected and the observed proportions deviate, then the assumed covariance between W_{CE}^1 and W_{CE}^{1+2} is no longer correct, and thus strict control of the global type I error rate cannot be guaranteed. Therefore, for time-to-event endpoints it is more realistic to choose an adaptive design, where the first- and second-stage test statistics are combined using predefined weights. The advantage of the latter approach is that the covariance then exclusively depends on the weights and is thus exactly known in the planning stage.

As shown by Wassmer (2006), an adaptive design for a time-to-event endpoint can be defined by means of the following test statistics for the interim and the final analysis, respectively

$$\widetilde{W}^1_{CE} := W^1_{CE},\ \widetilde{W}^{1+2}_{CE} := \frac{w_1 \cdot W^1_{CE} + w_2 \cdot \widetilde{W}^2_{CE}}{\sqrt{w_1^2 + w_2^2}} \sim N(0,1), \tag{6.12}$$

where the weights w_1, w_2 are predefined in the planning stage. The covariance between these test statistics now exclusively depends on the predefined weights w_1, w_2 and is thus known in advance

$$Cov\left(\widetilde{W}^1_{CE}, \widetilde{W}^{1+2}_{CE}\right) = \frac{w_1}{\sqrt{w_1^2 + w_2^2}}. \tag{6.13}$$

Hence, standard group-sequential boundaries can be applied as, for example, the Pocock boundaries (Pocock 1977) or the boundaries according to O'Brien and Fleming (1979). Choosing $w_1 = \sqrt{\psi}, w_2 = \sqrt{1-\psi}$ would correspond to an optimal weighting of the two stages in case the observed proportion of events at interim corresponds to the expected proportion, i.e., $\psi = \widehat{\psi}$. If during the interim analysis the actual information fraction at interim is equal to the one specified by the weights, this procedure is equivalent to the standard group-sequential approach. If the proportion of observed events at interim is smaller than the anticipated information fraction $\widehat{\psi}$—for example, due to an increase of the initially planned total number of expected events—then the second-stage data are down-weighted.

In case a single time-to-event endpoint is chosen as primary endpoint, the definition of a corresponding two-stage design is equivalent and will therefore be omitted.

6.4 Incorporating Stopping for Futility

The option to stop a trial early for futility has several relevant advantages. First, from an ethical viewpoint, fewer patients are treated with an ineffective drug which, as a second aspect, also saves financial and time resources. In the context of a composite endpoint with high planning uncertainties, the study should preferably be stopped early if the treatment effect is lower than anticipated. In general, stopping for futility is possible at any time and independent of any predefined rules without compromising the significance level as an early acceptance of the corresponding null hypothesis decreases the global probability to reject the null hypothesis. For given normally distributed test statistics Z^1 and Z^{1+2} and a predefined futility bound α_0, the test decision is given as follows. At the interim analysis, after recruitment of n_1

patients, the null hypothesis is rejected whenever

$$Z^1 \geq z_{1-\alpha_1}. \tag{6.14}$$

Additionally, the study is stopped for futility whenever

$$Z^1 \leq z_{1-\alpha_0}. \tag{6.15}$$

Otherwise, the trial is continued, and an additional number of n_2 patients are recruited. At the final analysis, the null hypothesis is rejected if

$$Z^{1+2} \geq z_{1-\alpha_{1+2}}. \tag{6.16}$$

In the context of group-sequential (adaptive) designs, it can generally be differentiated between binding and non-binding stopping for futility rules (Schüler et al. 2017; Bretz et al. 2009). Binding rules imply that stopping for futility at the interim analysis is obligatory whenever the futility criteria are met. The advantage of a binding futility rule is that the local significance levels might be increased in order to exhaust the global significance level. The local stage-wise levels $\alpha_1, \ \alpha_{1+2}$ can now be chosen such that

$$P_{H_0}\left(Z^1 \geq z_{1-\alpha_1} \vee \left(z_{1-\alpha_0} \leq Z^1 \leq z_{1-\alpha_1} \wedge Z^{1+2} \geq z_{1-\alpha_{1+2}}\right)\right) = \alpha. \tag{6.17}$$

In contrast, the non-binding version does not commit early futility stopping. Therefore, the local significance levels cannot be adjusted, and the global significance level is not fully exhausted. As explored by Schüler et al. (2017), in clinical trial applications, non-binding futility boundaries are usually applied because they allow reacting flexibly to interim results such as adverse events or new external information. However, quantifying the performance properties of non-binding rules is impossible as the study progress is not predictable from the observed effect at interim. Therefore, in order to judge the influence of futility rules on the design performance, binding rules must be investigated. The resulting performance properties of a binding rule can also be used to approximate the performance properties of more liberal non-binding rules as some deviations from the binding rule will not importantly affect the performance. For the reasons specified above, it seems reasonable to investigate binding futility stopping criteria at interim but without increasing the local significance levels.

Generally, incorporating stopping for futility allows to stop unpromising trials at an early stage which is of particular importance if the expected treatment effect is difficult to predict in the planning stage. Therefore, for clinical trials with a composite primary endpoint, this option is of particular relevance. On the other hand, stopping for futility also implies a power loss of the corresponding design which has to be carefully considered. A possible approach to counterbalance the loss in power against the advantage to stop an inefficient trial is presented in the next section.

6.4.1 Choice of Futility Boundaries

Choosing adequate futility boundaries can be a difficult challenge as unfavorable futility boundaries may have serious consequences with respect to the performance of the study design. In the case of "sharp" futility boundaries, for example, if the study is stopped for futility whenever the one-sided p-value is larger than 0.2 (or equivalently whenever the normally distributed test statistic is smaller or equal than $z_{1-0.2} = 0.84$), the overall power loss can be substantial, and the study might be stopped for futility in too many situations caused by only small but not relevant deviations from the planning assumptions. In such cases, the probability of "wrongly" stopping for futility is high. In the case of "liberal" futility boundaries, given, for example, as a lower bound for the p-value of 0.8 (or equivalently as an upper bound for the normally distributed test statistic of $z_{1-0.8} = -0.84$), the overall power loss is quite small, but at the same time small or adverse treatment effects often do not result in an early stopping as intended. Therefore, in this case the probability of "correctly" stopping for futility is low. In order to judge the influence of futility stopping on the overall study design performance, binding rules can be studied. Even if deviations from these binding rules will be necessary in practice, it is favorable to start with futility rules that are "optimized" for the binding case.

The concept of "optimized" futility boundaries presented in the following was originally published by Schüler et al. (2017). The idea is to keep the rate of "correctly" stopping for futility high and to simultaneously restrict the loss in power and the rate of "wrongly" stopping for futility. The expression "optimal" should be used with care in this context, as there exist many other optimality criteria for group-sequential and adaptive designs, such as the achieved power or the expected sample size under the null or under the alternative hypothesis. To provide optimal futility boundaries in the above sense, an "admissible condition" is defined as follows.

Definition 6.4.1 (Admissible Futility Boundary) For a given one-sided test problem, let η be the assumed treatment effect under the alternative hypothesis, for which the trial should be powered with power $1 - \beta \in [0; 1]$. Let $\beta_{loss} \in [0; 1]$ denote the maximally admissible overall power loss under an effect η in a two-stage group-sequential design with binding stopping for futility, and let $\pi_{wrong} \in [0; 1]$ denote the maximally admissible probability of stopping for futility under any effect within the alternative hypothesis, the so-called "wrongly" stopping for futility rate. Then a corresponding futility boundary α_f is called **admissible futility boundary** if the following conditions are met:

1. The probability to correctly reject the null hypothesis under a true treatment effect η is larger or equal to $1 - \beta - \beta_{loss}$,
2. The probability to wrongly stop for futility, that is, to accept the null hypothesis at interim, under a true treatment effect η is less or equal to π_{wrong}.

For predefined values of β_{loss} and π_{wrong}, there exist a number of admissible futility boundaries differing in the probability of "correctly" stopping for futility under small or adverse effects belonging to the null or alternative hypothesis. In order to determine optimal futility boundaries, the probability of early stopping for futility in case of a small or adverse effect which deviates considerably from the anticipated treatment effect η should be preferably high, which motivates the following definition.

Definition 6.4.2 (Optimal Futility Boundary for Correct Stopping) For a given one-sided test problem, let η be the assumed treatment effect under the alternative hypothesis, for which the trial should be powered with power $1 - \beta \in [0; 1]$. Let $A_{\beta_{loss}, \pi_{wrong}}$ denote the set of all the admissible futility boundaries for a maximally admissible power loss $\beta_{loss} \in [0; 1]$ and a maximally admissible "wrongly" stopping for futility rate $\pi_{wrong} \in [0; 1]$. Let $\eta_{stop} < \eta$ denote the largest effect favoring the alternative hypothesis for which stopping the study for futility would still be considered as "correct". Then the maximal admissible futility boundary $\alpha_{opt} \in A_{\beta_{loss}, \pi_{wrong}}$ for which the study is still "correctly" stopped for futility under the effect η_{stop} is called **optimal futility boundary for correct stopping (OCS futility boundary)**.

Note that all futility boundaries that are larger than the OCS boundary automatically meet the admissible conditions. By applying an OCS futility boundary, it is possible to already define in the planning stage when a trial is no longer worth to be continued after the interim stage. By this, at the final analysis it is very unlikely to observe a small or an adverse treatment effect which prevents from interpretation problems. An algorithm to provide OCS futility boundaries for a predefined group-sequential design is provided in Chap. 7.

6.5 Sample Size Calculation

For group-sequential and adaptive designs, sample size calculation can be performed using the software `ADDPLAN` (ICON plc 2014) which comprises four different modules. Sample size calculation for a single endpoint can be realized by means of the module `ADDPLAN base`. In the following, the underlying mathematical considerations yielding the required sample size will be deduced.

6.5.1 Sample Size Calculation for Group-Sequential Designs

In a two-stage group-sequential design, sample size calculation requires, in addition to the parameters for the corresponding one-stage design, specification of the employed stage-wise local levels α_1, α_{1+2} and, if applicable, a binding futility boundary α_0 to be applied at the interim analysis. Moreover, the information fraction ψ at interim must be specified. The first- and the second-stage sample sizes are then chosen such that for a given effect η under the alternative hypothesis it holds

$$P_{\eta \in H_1}\left(Z^1 \geq z_{1-\alpha_1} \vee \left(z_{1-\alpha_0} \leq Z^1 \leq z_{1-\alpha_1} \wedge Z^{1+2} \geq z_{1-\alpha_{1+2}}\right)\right) \geq 1-\beta. \quad (6.18)$$

6.5.2 *Sample Size Recalculation for Adaptive Designs*

In general, adaptive designs allow for a variety of design modifications (Irle and Schäfer 2012), e.g., a change in the trial population (Mehta et al. 2014; Friede et al. 2012) or the treatment arms (Carreras et al. 2015; Di Scala and Glimm 2011; Stallard and Todd 2003) thereby still guaranteeing control of the global type I error rate. In particular, within an adaptive design it is possible to change the second-stage sample size at interim freely according to an arbitrary rule (Lehmacher and Wassmer 1999; Bauer and Köhne 1994). A commonly applied strategy is to recalculate the sample size $n_{2,recalc}$ such that the conditional power based on the assumption that the observed value of the test statistic at interim t^1_{CE} corresponds to the true effect, is sufficiently high. For a composite binary endpoint, this means

$$P_{t^1_{CE}}\left(\widetilde{T}^{1+2}_{CE} \geq z_{1-\alpha_{1+2}}\right) = 1-\Phi\left(z_{1-\alpha_{1+2}} \cdot \frac{\sqrt{w_1^2+w_2^2}}{w_2} - t^1_{CE} \cdot \left(\frac{w_1}{w_2} + \sqrt{\frac{n_{2,recalc}}{n_1}}\right)\right)$$
$$\geq 1-\beta.$$

Thus, the required sample size for the second stage is given by the smallest integer fulfilling

$$n_{2,recalc} \geq n_1 \cdot \left(\frac{-z_{1-\beta} - z_{1-\alpha_{1+2}} \cdot \frac{\sqrt{w_1^2+w_2^2}}{w_2}}{t^1_{CE}} + \frac{w_1}{w_2}\right)^2. \quad (6.19)$$

For a composite time-to-first-event endpoint, the inequality for the conditional power to be solved is given by

$$P_{w^1_{CE}}\left(\widetilde{W}^{1+2}_{CE} \geq z_{1-\alpha_{1+2}}\right) = 1-\Phi\left(z_{1-\alpha_{1+2}} \cdot \frac{\sqrt{w_1^2+w_2^2}}{w_2} - w^1_{CE} \cdot \left(\frac{w_1}{w_2} + \sqrt{\frac{d_{2,recalc}}{d_1}}\right)\right)$$
$$\geq 1-\beta,$$

where again w^1_{CE} is the observed value of the test statistic at interim. Thus, the required number of events for the second-stage $d_{2,recalc}$ is given by the smallest integer fulfilling

$$d_{2,recalc} \geq d_1 \cdot \left(\frac{-z_{1-\beta} - z_{1-\alpha_{1+2}} \cdot \frac{\sqrt{w_1^2+w_2^2}}{w_2}}{w_{CE}^1} + \frac{w_1}{w_2} \right)^2. \tag{6.20}$$

References

Bauer, P., & Köhne, K. (1994). Evaluation of experiments with adaptive interim analyses. *Biometrics, 50*, 1029–1041.

Bretz, F., Koenig, F., Brannath, W., Glimm, E., & Posch, M. (2009). Adaptive designs for confirmatory clinical trials. *Statistics in Medicine, 28*, 1181–1217.

Brückner, M., & Brannath, W. (2016). Sequential tests for non-proportional hazards data. *Lifetime Data Analysis, 23*, 1–14.

Carreras, M., Gutjahr, G., & Brannath, W. (2015). Adaptive seamless designs with interim treatment selection: A case study in oncology. *Statistics in Medicine, 34*, 1317–1333.

Di Scala, L., & Glimm, E. (2011). Time-to-event analysis with treatment arm selection at interim. *Statistics in Medicine, 30*, 3067–3081.

Friede, T., Parsons, N., & Stallard, N. (2012). A conditional error function approach for subgroup selection in adaptive clinical trials. *Statistics in Medicine, 31*, 4309–4320.

Hwang, I. K., Shih, W. J., & De Cani, J. S. (1990). Group sequential designs using a family of type I error probability spending functions. *Statistics in Medicine, 9*, 1439–1445.

ICON plc. ADDPLAN Software. (2014). Version 6.1.1. http://www.iconplc.com/technology/addplan/. Accessed April 9, 2018.

Irle, S., & Schäfer, H. (2012). Interim design modifications in time-to-event studies. *Journal of the American Statistical Association, 107*, 341–348.

Jennison, C., & Turnbull, B. W. (1999). *Group sequential methods with applications to clinical trials*. Boca Raton: CRC.

Kim, K., & Demets, D. L. (1987). Design and analysis of group sequential tests based on the type I error spending rate function. *Biometrika, 74*, 149–154.

Lan, K. G., & DeMets, D. L. (1983). Discrete sequential boundaries for clinical trials. *Biometrika, 70*, 659–663.

Lehmacher, W., & Wassmer, G. (1999). Adaptive sample size calculations in group sequential trials. *Biometrics, 55*, 1286–1290.

Mehta, C., Schäfer, H., Daniel, H., & Irle, S. (2014). Biomarker driven population enrichment for adaptive oncology trials with time to event endpoints. *Statistics in Medicine, 33*, 4515–4531.

O'Brien, P. C., & Fleming, T. R. (1979). A multiple testing procedure for clinical trials. *Biometrics, 35*, 549–556.

Pocock, S. J. (1977). Group sequential methods in the design and analysis of clinical trials. *Biometrika, 64*, 191–199.

Schüler, S., Kieser, M., & Rauch, G. (2017). Choice of futility boundaries for group sequential designs with two endpoints. *BMC Medical Research Methodology, 17*, 119.

Stallard, N. L., & Todd, S. (2003). Sequential designs for phase III clinical trials incorporating treatment selection. *Statistics in Medicine, 22*, 689–703.

Wassmer, G. (1999). Statistische Testverfahren für gruppensequentielle und adaptive Pläne in klinischen Studien: theoretische Konzepte und deren praktische Umsetzung mit SAS, Mönch.

Wassmer, G. (2006). Planning and analyzing adaptive group sequential survival trials. *Biometrical Journal, 48*, 714–729.

Chapter 7
Related Software Code

This chapter provides exemplary source code written in the software `R` implementing different approaches outlined in Part II of this book.

7.1 Sample Size Calculation Based on Expected Power for a Binary Endpoint

```
library(mvtnorm)
library(triangle)

####################################################
# Parameters to specify                            #
####################################################
#
# nI       = Sample size of intervention group
# gamma    = nI/nC
# pC       = Event rate in the control group
# pI       = Event rate in the intervention group
# delta    = Non-inferiority margin, <= 0
# alpha    = One-sided significance level
# nsim     = Number of simulations for Monte-Carlo
#            integration to calculate expected power
# nI_start = Starting value for sample size of
#            intervention group
# power    = Desired power
#
# In addition prior distributions must be
# prespecified in "expected_Powerfunction"!!!!!!
####################################################


#####################################################
# Program to calculate the frequentist power for    #
# a superiority or non-inferiority test problem to  #
# compare rates between two groups                  #
#####################################################
```


G. Rauch et al., *Planning and Analyzing Clinical Trials with Composite Endpoints*, Springer Series in Pharmaceutical Statistics,
https://doi.org/10.1007/978-3-319-73770-6_7

```
power=function(nI, gamma, pC, pI, delta, alpha){

  nC = nI/gamma

  #--------------------------------------------------#
  # Calculation of the superiority test statistic   #
  #--------------------------------------------------#

  if(delta==0){
    if (pC==0 & pI==0){
      t=0
    }else{
p = (pI*nI+pC*nC)/(nI+nC)
# Variance under no effect (5.5)
Var_0 = p*(1-p)*((nI+nC)/(nI*nC))
# Variance for given treatment effect
Var = pC*(1-pC)/nC+pI*(1-pI)/nI
# Normal approximation test for rates (5.3)
t = (pC-pI)/(sqrt(Var_0))
#Transformed critical value for power calculation
help = (qnorm(1-alpha)*sqrt(Var_0)
-(pC-pI))/sqrt(Var)
    }
  }

#----------------------------------------------------#
# Calculation of the non-inferiority test statistic #
# as described in Farrington and Manning 1990.      #
# The restricted maximum likelihood estimator pI_far#
# under the null hypothesis restriction pC-pI=delta #
# is given as the unique solutions in (-delta,1) of #
# the maximum likelihood equation: ax^3+bx^2+cx+d=0 #
#----------------------------------------------------#

  if(delta<0){
    theta   = 1/gamma
    s0      = -delta
    a       = 1+theta
    b       = -(1+theta+pI+theta*pC+s0*(theta+2))
    c       = s0^2+s0*(2*pI+theta+1)+pI+theta*pC
    d       = -1*pI*s0*(1+s0)
    # ---------------------------------------------
    v       = (b**3)/((3*a)**3)-b*c/(6*a*a)+d/(2*a)
    u       = sign(v)*sqrt( b*b/((3*a)**2)-c/(3*a))
    w       = (1/3)*(pi+acos(v/(u**3)))
    # ---------------------------------------------
    pI_far = 2*u*cos(w) - b/(3*a)
    pC_far = pI_far-s0

# Variance under null effect delta (5.6)
Var_far=(pI_far*(1-pI_far)
 +pC_far*(1-pC_far)*gamma)/nI
```

```
# Variance for given treatment effect
Var    = pC*(1-pC)/nC+pI*(1-pI)/nI
# Non-inferiority test statistic for rates (5.4)
t      = (pC-pI-delta)/(sqrt(Var_far))
# Transformed critical value for power calculation
help   = (qnorm(1-alpha)*sqrt(Var_far)
-(pC-pI-delta))/sqrt(Var)
}

  #-----------------------------------------------#
  # Calculation of the frequentist power          #
  #-----------------------------------------------#

  pow  = 1-pnorm(help)

  return(pow)
}



#*********************************#
# Exemplary Calls                  #
#*********************************#

print("Frequentist Power")
power(100, 1, 0.4, 0.2, 0, 0.025)
power(100, 2, 0.4, 0.2, -0.05, 0.025)



#####################################################
# Calculation of the expected power                 #
# for different prior distributions                 #
# by Monte-Carlo integration                        #
#####################################################

expected_Power=function(nI,gamma,delta,alpha,nsim){

  pow = 0

  for(i in 1:nsim){
    #-----------------------------------------------#
    # Please specify prior distributions            #
    # for the event rates.                          #
    # Uniform and triangular priors                 #
    # fitting to the exemplary parameter            #
    # constellation provided as exemplary priors    #
    # (please comment/uncomment accordingly).       #
    #-----------------------------------------------#

    #-----------------------------------------------#
    # Exemplary priors 1: Uniform distribution      #
    #-----------------------------------------------#
```

```
    pC = runif(1,0.3,0.5)
    pI = runif(1,0.1,0.3)

    #-----------------------------------------------#
    # Exemplary priors 2: Triangular distribution   #
    #-----------------------------------------------#
    #pC = rtriangle(1,0.3,0.5,0.4)
    #pI = rtriangle(1,0.1,0.3,0.2)

    pow = pow+power(nI,gamma,pC,pI,delta,alpha)/nsim
  }
  return(pow)
}


#************************************#
# Exemplary calls                    #
#************************************#

print("Expected Power")
expected_Power(100,1,0,0.025,10000)
expected_Power(100,2,-0.05,0.025,10000)

#####################################################
# Calculation of the total sample size for          #
# predefined expected power                         #
#####################################################


sample_size=function(nI_start,power,
                     gamma,delta,alpha,nsim){

  nI  = nI_start
  pow = expected_Power(nI,gamma,delta,alpha,nsim)

  if (pow>power){
    print("Choose smaller starting value nI")
  }

  if(pow<power){

    while(pow<power){

      nI =nI+1
      pow=expected_Power(nI,gamma,delta,alpha,nsim)

    }

    nC=nI/gamma
    return(c(nI, nC,pow))
  }

}
```

```
#***********************************#
# Exemplary calls                   #
#***********************************#

print("Sample Size for predefined expected power")
sample_size(70,0.8,1,0,0.025,10000)
sample_size(70,0.8,2,-0.05,0.025,10000)
sample_size(100,0.8,2,-0.05,0.025,10000)
```

7.2 Sample Size Calculation for the Average Hazard Ratio

```
library(compiler)
library(testthat)
library(AHR)
library(survival)
library(coxphw)

####################################################
# Parameters to specify                            #
####################################################
#                                                  #
# lI    = Scale parameter of Weibull distribution  #
#         in the intervention group                #
# kI    = Shape parameter of Weibull distribution  #
#         in the intervention group                #
# lC    = Scale parameter of Weibull distribution  #
#         in the control group                     #
# kC    = Shape parameter of Weibull distribution  #
#         in the control group                     #
# nI    = Sample size of the intervention group    #
# gamma= nI/nC                                     #
# tac   = Accrual duration                         #
# tfu   = Minimal follow-up                        #
# nsim = Number of data sets simulations           #
####################################################


####################################################
# Simulating data                                  #
# for predefined Weibull event time distributions #
####################################################

data=function(lI,kI,lC,kC,nI,gamma,tac,tfu){

  group_I = c()
  group_C = c()

  T_I = c()
  T_C = c()
```

```
  C_I = c()
  C_C = c()

  nC = nI/gamma

  cens_I = tac*runif(nI,0,1)+tfu
  cens_C = tac*runif(nC,0,1)+tfu

  cause_I = c()
  cause_C = c()


  for(i in 1:nI){
    group_I[i]=1
    T_I[i]= rweibull(1, kI, scale = 1/lI)
    if (T_I[i]>=cens_I[i]){
      cause_I[i]=0
      T_I[i]=cens_I[i]
    }
    else if (T_I[i]<cens_I[i]){
      cause_I[i]=1
    }
  }

  for(i in 1:nC){
    group_C[i]=0
    T_C[i]= rweibull(1, kC, scale = 1/lC)
    if (T_C[i]>=cens_C[i]){
      cause_C[i]=0
      T_C[i]=cens_C[i]
    }
    else if (T_C[i]<cens_C[i]){
      cause_C[i]=1
    }
  }

  data=data.frame(id=1:(nI+nC),
            times=c(T_I,T_C),
            status=c(cause_I, cause_C),
            group=as.factor(c(group_I,group_C)))


  return(data)

}

data = cmpfun(data)


#####################################################
# Calculation of the power                          #
# for the average hazard ratio with gamma=1         #
# for predefined Weibull event time distributions #
#####################################################
```

```
power_ahr=function(lI,kI,lC,kC,nI,gamma,
                   tac,tfu,alpha,nsim){

  ahr           = numeric(nsim)
  power_ahr     = numeric(nsim)
  p_value_theta = numeric(nsim)

  for(i in 1:nsim){

    data=data(lI,kI,lC,kC,nI,gamma, tac, tfu)

    data$group = as.integer(data$group)
    data$times[data$times<0.00001] = 0.00001

    #--------------------------------------------#
    # Calculate AHR according to                 #
    # Kalbfleisch and Prentice with gamma=1      #
    #--------------------------------------------#

    fit=ahrWKM(5, Surv(times,status)~group,data,
           null.theta = NULL, contrast = NULL,
           multi.test = FALSE, bootstrap = 0,
           cov = TRUE, alpha = 1)

    ahr[i]<-1/fit$hr
    p_value_theta[i]=pnorm(fit$Z.loghr)

    if (p_value_theta[i]<=alpha){
      power_ahr[i]=1
    }
  }
  return(c(mean(ahr),mean(power_ahr)))
}

power_ahr = cmpfun(power_ahr)

#************************************#
# Exemplary calls                    #
#************************************#

print("AHR and related Power")
power_ahr(0.5,1.2,1,1,40,1,2,3,0.025,1000)
power_ahr(0.5,1.2,1,1,40,2,2,3,0.025,1000)

##################################################
# Calculation of required sample size for        #
# the power of the average hazard ratio with     #
# gamma=1 for predefined Weibull event time      #
# distributions.                                 #
##################################################

sample_size_ahr=function(lI,kI,lC,kC,nI_start,
gamma,tac,tfu,alpha,nsim,power){
```

```
#-----------------------------------------------#
# Plot underlying event time distributions      #
#-----------------------------------------------#

par(mfrow=c(1,3))
plot(function(x) exp(-(lI*x)**kI),0,5,lwd=2,lty=1,
     ylab="Probability to be event-free",
     xlab="Time", ylim=c(0, 1))
par(new=TRUE)
plot(function(x) exp(-(lC*x)**kC),0,5,lwd=2,lty=2,
     ylab="Probability to be event-free",
     xlab="Time",ylim=c(0, 1))
legend(1,1,c("Intervention", "Control"),
       lty = c(1,2),lwd=c(2,2),bty="n")
plot(function(x) (kI*lI*(lI*x)**(kI-1)),0,5,
     lwd=2,lty=1,ylab="Hazard",
     xlab="Time",ylim=c(0,5))
par(new=TRUE)
plot(function(x) (kC*lC*(lC*x)**(kC-1)),0,5,
     lwd=2,lty=2,ylab="Hazard",
     xlab="Time",  ylim=c(0,5))
legend(1,5,c("Intervention", "Control"),
       lty = c(1,2),lwd=c(2,2),bty="n")
plot(function(x)
   (kC*lC*(lC*x)**(kC-1))/(kI*lI*(lI*x)**(kI-1)),
   0,5, lwd=2,lty=1,ylab="Hazard Ratio",
   xlab="Time",ylim=c(0,5))

#-----------------------------------------------#

nI  = nI_start
pow = power_ahr(lI,kI,lC,kC,nI,gamma,
                tac,tfu,alpha,nsim)[2]

if (pow>power){
  print("Choose smaller starting value nI")
}

if(pow<power){

  while(pow<power){

    nI =nI+gamma
    pow = power_ahr(lI,kI,lC,kC,nI,gamma,
                    tac,tfu,alpha,nsim)[2]

  }

  nC=nI/gamma
  return(c(nI,nC,pow))
}

}
```

```
#*************************************#
# Exemplary calls                     #
#*************************************#

print("Sample size for AHR (n_I, n_C, power)")
sample_size_ahr(0.5,1.2,1,1,40,1,2,3,0.025,1000,0.8)
sample_size_ahr(0.5,1.2,1,1,40,2,2,3,0.025,1000,0.8)
```

7.3 Calculation of the OCS Futility Boundary

7.3.1 OCS Futility Boundary for a Binary Endpoint

```
library(mvtnorm)

###################################################
# Calculation of the OCS futility boundary        #
# for a binary endpoint.                          #
# Interim analysis is performed after a fixed     #
# number of patients n_1=infotime*n_12            #
###################################################



###################################################
# "optimal_pow": Help function searching for      #
# futility boundaries which fullfil               #
# the power criteria given in Definition 6.4.1    #
###################################################

###################################################
# Parameters to specify                           #
###################################################
#
# alpha_1    = Group-sequential one-sided local
#              significance level (Pocock)
# infotime   = Information rate at interim
# beta       = Overall type II error rate;
#              (1-beta) anticipated overall power
#              under effect eta
# beta_loss  = Maximal admissible overall power loss,
#              see Definition 6.4.1,Sect. 6.4.1.
# T_1        = Value of the interim test statistic
# T_12       = Value of the overall test statistic
#              at the final analysis
###################################################
```

```
optimal_pow <- function(alpha_1, infotime, beta,
                        beta_loss, T_1, T_12){
x = T_1
y = T_12

#---------------------------------------------#
# Specify starting values                     #
# for power and futility boundary             #
#---------------------------------------------#
pow     = 0.5
alpha_f = 0.05

while((pow < (1 - beta - beta_loss))&
      (alpha_f < 0.99)){
     alpha_f <- alpha_f + 0.01
     pow_1 <- 1-pnorm(qnorm(1-alpha_1), x, 1)
     cov_mtx <- matrix(c(1,sqrt(infotime),
                    sqrt(infotime),1),ncol=2)
     pow_2 <- pmvnorm(lower=c(qnorm(1-alpha_f),
                    qnorm(1-alpha_1)),
                    upper=c(qnorm(1-alpha_1),Inf),
                    mean=c(x,y), sigma=cov_mtx)
     pow <- pow_1 + pow_2
}

  out <- c(pow,alpha_f)
  return(out)

}

##################################################
# "optimal_wfut": Help function searching for    #
# futility boundaries which limit                #
# the probability of wrongly stopping for        #
# futility compare Definition 6.4.1              #
##################################################


##################################################
# Parameters to specify                          #
##################################################
#
# pi_wrong  = Maximal admissible rate of
#             wrongly stopping for futility ,
#             see Definition 6.4.1, Sect. 6.4.1.
# T_1       = Value of the interim test statistic
##################################################

optimal_wfut <- function(pi_wrong, T_1){

  #---------------------------------------------#
  # Specify the probability to stop for         #
  # futility and for the futility boundary      #
  #---------------------------------------------#
```

```
  wfut    = 1
  alpha_f = 0.01

  while(wfut > pi_wrong){
    alpha_f <- alpha_f + 0.01
    wfut <- pnorm(qnorm(1-alpha_f), T_1, 1)
  }

  out <- c(wfut,alpha_f)
  return(out)

}


###################################################
###################################################
# "optimal_bound": Main function for calculation  #
# of OCS futility boundary                        #
###################################################
###################################################

###################################################
# Parameters to specify                           #
###################################################
#
# pC        = Assumed event rate in the control group
#             for the composite endpoint
# pI        = Assumed event rate in the intervention
#             group for the composite endpoint
# n_12      = Total patient number at the
#             final analysis assuming the effect
#             of eta
# alpha_l   = Group-sequential one-sided local
#             significance level (Pocock)
# infotime  = Information rate at interim
# beta_loss = Maximal admissible overall power loss,
#             see Definition 6.4.1, Sect. 6.4.1.
# pi_wrong  = Maximal admissible rate of
#             wrongly stopping for futility,
#             see Definition 6.4.1, Sect. 6.4.1.
# pI_stop   = eta_stop=pC-pI_stop correspond to
#             largest effect for
#             which stopping for futility would
#             still be considered as correct in
#             terms of the absolute risk difference,
#             see Definition 6.4.2, Sect. 6.4.1.
###################################################

optimal_bound <- function(pC, pI, n_12, alpha_l,
                          infotime, beta_loss,
                          pi_wrong, pI_stop){
```

```
#------------------------------------------------#
# Calculation of the stage wise test statistics #
# according to                                   #
# Sect. 5.1.2. Eq. (5.10)                        #
# under original planning assumptions eta        #
# assuming equal sample sizes per group          #
#------------------------------------------------#

eta <-pC-pI
p<-(pC+pI)/2
T_1 <- eta/sqrt(p*(1-p)*(4/(n_12*infotime)))
T_12 <- eta/sqrt(p*(1-p)*(4/(n_12)))


#------------------------------------------------#
#Power without futility stop                     #
#------------------------------------------------#

pow_1 <- 1- pnorm(qnorm(1-alpha_1), T_1, 1)
cov_mtx <- matrix(c(1,sqrt(infotime),
                    sqrt(infotime),1),ncol=2)
pow_2 <- pmvnorm(lower=c(-Inf,qnorm(1-alpha_1)),
                 upper=c(qnorm(1-alpha_1),Inf),
                 mean=c(T_1,T_12),
                 sigma=cov_mtx)
pow <- pow_1 + pow_2
beta<-1-pow[1]

#------------------------------------------------#
# Calculation of optimal futility boundary       #
# alpha_opt                                      #
#------------------------------------------------#


pow_o  <- optimal_pow(alpha_1, infotime, beta,
                      beta_loss, T_1, T_12)
wfut_o <- optimal_wfut(pi_wrong, T_1)

alpha_opt <- max(wfut_o[2], pow_o[2])

#------------------------------------------------#
# Determination of power, probability of         #
# wrongly stopping and probability of correctly #
# stoppping (unter eta_stop and H_0)             #
#------------------------------------------------#

#------------------------------------------------#
#Power using optimal futility boundary alpha_opt#
#------------------------------------------------#
pow_1 <- 1- pnorm(qnorm(1-alpha_1), T_1, 1)
cov_mtx <- matrix(c(1,sqrt(infotime),
                    sqrt(infotime),1),ncol=2)
pow_2 <- pmvnorm(lower=c(qnorm(1-alpha_opt),
                         qnorm(1-alpha_1)),
```

```
                     upper=c(qnorm(1-alpha_1),Inf),
                     mean=c(T_1,T_12),
                     sigma=cov_mtx)
    op_pow <- pow_1 + pow_2


    #-----------------------------------------------#
    # Probability of wrongly stopping for futility  #
    # using optimal futility boundary alpha_opt     #
    #-----------------------------------------------#

    op_wfut <- pnorm(qnorm(1-alpha_opt), T_1, 1)

    #-----------------------------------------------#
    # Probability of correctly stopping for futility#
    # using optimal futility boundary alpha_opt     #
    # (based on eta_stop)                           #
    #-----------------------------------------------#

    eta_stop<-pC-pI_stop
    p<-(pC+pI_stop)/2
    z <- (eta_stop)/sqrt(p*(1-p)*(4/(n_12*infotime)))
    op_cfut_eta_stop <-pnorm(qnorm(1-alpha_opt),z,1)

    #-----------------------------------------------#
    # Probability of correctly stopping for futility#
    # using optimal futility boundary alpha_opt     #
    # (based on H_0)                                #
    #-----------------------------------------------#
    op_cfut_H0 <- pnorm(qnorm(1-alpha_opt), 0, 1)
    z_op <- qnorm(1-alpha_opt)

    print("OCS futility boundary")
    print(alpha_opt)
    print("OCS futility z-quantile")
    print(z_op)
    print("Power without futility stop")
    print(pow[1])
    print("Power with futility stop")
    print(op_pow[1])
    print("Prob. of wrongly stopping for fuitility")
    print(op_wfut)
    print("Prob. of correctly stopping for fuitility")
    print("...under effect of eta_stop")
    print(op_cfut_eta_stop)
    print("Prob. of correctly stopping for fuitility")
    print("...under H_0")
    print(op_cfut_H0)

}
```

```
#************************************#
# Exemplary Calls                    #
#************************************#

#######################################
# Specify                             #
# pC, pI, n_12, alpha_1, infotime,    #
# beta_loss, pi_wrong, pI_stop        #
#######################################

optimal_bound(0.3,0.1,138,0.0147,0.5,
              0.001,0.025,0.25)
```

7.3.2 OCS Futility Boundary for a Time-to-Event Endpoint

```
library(mvtnorm)

#####################################################
# Calculation of the OCS futility boundary          #
# for a time-to-event endpoint .                    #
# Interim analysis is performed after a fixed       #
# number of patients n1                             #
#####################################################




#####################################################
# "optimal_pow": Help function searching for        #
# futility boundaries which fullfil                 #
# the power criteria given in Definition 6.4.1      #
#####################################################


#####################################################
# Parameters to specify                             #
#####################################################
#
# alpha_1   = Group-sequential one-sided local
#             significance level (Pocock)
# infotime  = Information rate at interim assuming
#             the effect of eta
# beta      = Overall type II error rate;
#             (1-beta) anticipated overall power
# beta_loss = Maximal admissible overall power loss,
#             see Definition 6.4.1,Sect. 6.4.1.
# W_1       = Value of the interim test statistic
# W_12      = Value of the overall test statistic
#             at the final analysis
#####################################################
```

```
optimal_pow <- function(alpha_1, infotime, beta,
                        beta_loss, W_1, W_12){
x = W_1
y = W_12

#-------------------------------------------#
# Specify starting values                   #
# for power and futility boundary           #
#-------------------------------------------#
pow     = 0.5
alpha_f = 0.05

while((pow < (1 - beta - beta_loss))&
     (alpha_f < 0.99)){
     alpha_f <- alpha_f + 0.01
     pow_1 <- 1-pnorm(qnorm(1-alpha_1), x, 1)
     cov_mtx <- matrix(c(1,sqrt(infotime),
                    sqrt(infotime),1),ncol=2)
     pow_2 <- pmvnorm(lower=c(qnorm(1-alpha_f),
                    qnorm(1-alpha_1)),
                    upper=c(qnorm(1-alpha_1),Inf),
                    mean=c(x,y), sigma=cov_mtx)
     pow <- pow_1 + pow_2
}

  out <- c(pow,alpha_f)
  return(out)

}

###################################################
# "optimal_wfut": Help function searching for     #
# futility boundaries which limit                 #
# the probability of wrongly stopping for         #
# futility compare Definition 6.4.1               #
###################################################


###################################################
# Parameters to specify                           #
###################################################
#
# pi_wrong  = Maximal admissible rate of
#             wrongly stopping for futility ,
#             see Definition 6.4.1, Sect. 6.4.1.
# W_1       = Value of the interim test statistic
###################################################

optimal_wfut <- function(pi_wrong, W_1){

  #-------------------------------------------#
  # Specify the probability to stop for       #
  # futility and for the futility boundary    #
  #-------------------------------------------#
```

```
  wfut    = 1
  alpha_f = 0.01

  while(wfut > pi_wrong){
    alpha_f <- alpha_f + 0.01
    wfut <- pnorm(qnorm(1-alpha_f), W_1, 1)
  }

  out <- c(wfut,alpha_f)
  return(out)

}

###################################################
###################################################
# "optimal_bound": Main function for calculation  #
# of OCS futility boundary                        #
###################################################
###################################################

###################################################
# Parameters to specify                           #
###################################################
#
# eta        = haz^C/haz^I, assumed effect in terms
#              of the hazard ratio for
#              the composite endpoint
# d_12       = Total number of expected events at the
#              final analysis assuming the effect
#              of eta
# alpha_l    = Group-sequential one-sided local
#              significance level (Pocock)
# infotime   = Information rate at interim
# beta       = Overall type II error rate;
#              (1-beta) desired overall power
# beta_loss = Maximal admissible overal power loss,
#              see Definition 6.4.1, Sect. 6.4.1.
# pi_wrong   = Maximal admissible rate of
#              wrongly stopping for futility,
#              see Definition 6.4.1, Sect. 6.4.1.
# eta_stop   = haz^C/haz^I_stop, largest effect for
#              which stopping for futility would still
#              be considered as correct in terms
#              of the hazard ratio,
#              see Definition 6.4.2, Sect. 6.4.1.
# d_12_stop = total number of expected events at the
#              final analysis assuming the effect
#              of eta_stop for required sample size
#              under eta
#--------------------------------------------------
# Remark: The number of expected events at the final
#         analysis d_12 and d_12_stop are
#         approximated by using the Simpson'rule,
#         compare Eq. (5.27), (5.28), (5.29)
```

```
#         in Section 5.2.3.1)
####################################################

optimal_bound <- function(eta, d_12, alpha_1,
                          infotime, beta, beta_loss,
                          pi_wrong, eta_stop,
                          d_12_stop){

  #------------------------------------------------#
  # Calculation of the stage wise test statistics #
  # according to                                   #
  # Sect. 5.2.2. Eq. (5.21)                        #
  # under original planning assumptions eta        #
  # assuming equal sample sizes per group.         #
  #------------------------------------------------#
  # d_12*infotime corresponds to the number of     #
  # events at interim under the assumed effect eta#
  # after inclusion of n1 patients                 #
  #------------------------------------------------#

  W_1 <- log(eta)*sqrt(d_12*infotime)/2
  W_2 <- log(eta)*sqrt(d_12)/2

  #------------------------------------------------#
  # Determine incremental test statistic           #
  # for stage 2, compare Sect. 6.3. Eq. (6.10)     #
  # assuming equal sample sizes per group          #
  #------------------------------------------------#

  W_2_incr <- ((sqrt(d_12)*W_2-
                sqrt(d_12*infotime)*W_1)/
                sqrt(d_12-d_12*infotime))

  #------------------------------------------------#
  # Determine of the overall test statistic,       #
  # compare Sect. 6.3. Eq. (6.10)                  #
  #------------------------------------------------#

  W_12 <- sqrt(infotime)*W_1 +
          sqrt(1-infotime)*W_2_incr

  #------------------------------------------------#
  # Calculation of optimal futility boundary       #
  # alpha_opt                                      #
  #------------------------------------------------#


  pow_o  <- optimal_pow(alpha_1, infotime, beta,
                        beta_loss, W_1, W_12)
  wfut_o <- optimal_wfut(pi_wrong, W_1)

  alpha_opt <- max(wfut_o[2], pow_o[2])
```

```
#--------------------------------------------------#
# Determination of power, probability of           #
# wrongly stopping and probability of correctly #
# stoppping (unter eta_stop and H_0)               #
#--------------------------------------------------#

#--------------------------------------------------#
#Power using optimal futility boundary alpha_opt#
#--------------------------------------------------#
pow_1 <- 1- pnorm(qnorm(1-alpha_1), W_1, 1)
cov_mtx <- matrix(c(1,sqrt(infotime),
                    sqrt(infotime),1),ncol=2)
pow_2 <- pmvnorm(lower=c(qnorm(1-alpha_opt),
                         qnorm(1-alpha_1)),
                 upper=c(qnorm(1-alpha_1),Inf),
                 mean=c(W_1,W_12),
                 sigma=cov_mtx)
op_pow <- pow_1 + pow_2

#--------------------------------------------------#
#Power without futility stop                       #
#--------------------------------------------------#
pow_1 <- 1- pnorm(qnorm(1-alpha_1), W_1, 1)
cov_mtx <- matrix(c(1,sqrt(infotime),
                    sqrt(infotime),1),ncol=2)
pow_2 <- pmvnorm(lower=c(-Inf,
                         qnorm(1-alpha_1)),
                 upper=c(qnorm(1-alpha_1),Inf),
                 mean=c(W_1,W_12),
                 sigma=cov_mtx)
pow <- pow_1 + pow_2

#--------------------------------------------------#
# Probability of wrongly stopping for futility  #
# using optimal futility boundary alpha_opt     #
#--------------------------------------------------#

op_wfut <- pnorm(qnorm(1-alpha_opt), W_1, 1)

#--------------------------------------------------#
# Probability of correctly stopping for futility#
# using optimal futility boundary alpha_opt     #
# (based on eta_stop)                           #
#--------------------------------------------------#
# d_12_stop*infotime corresponds to the number  #
# of events at interim under the effect eta_stop#
# after inclusion of n1 patients                #
#--------------------------------------------------#

z <- log(eta_stop)*sqrt(d_12_stop*infotime)/2
op_cfut_eta_stop <-pnorm(qnorm(1-alpha_opt),z,1)
```

```
  #-------------------------------------------------#
  # Probability of correctly stopping for futility#
  # using optimal futility boundary alpha_opt     #
  # (based on H_0)                                #
  #-------------------------------------------------#
  op_cfut_H0 <- pnorm(qnorm(1-alpha_opt), 0, 1)
  z_op <- qnorm(1-alpha_opt)

  print("OCS futility boundary")
  print(alpha_opt)
  print("OCS futility z-quantile")
  print(z_op)
  print("Power without futility stop")
  print(pow[1])
  print("Power with futility stop")
  print(op_pow[1])
  print("Prob. of wrongly stopping for fuitility")
  print(op_wfut)
  print("Prob. of correctly stopping for fuitility")
  print("...under effect of eta_stop")
  print(op_cfut_eta_stop)
  print("Prob. of correctly stopping for fuitility")
  print("...under H_0")
  print(op_cfut_H0)

}


#**************************************#
# Exemplary Calls                      #
#**************************************#

########################################
# Specify                              #
# eta, d_12, alpha_1, infotime,        #
# beta_loss, pi_wrong, eta_stop,       #
# d_12_stop                            #
########################################

#----------------------------------------------#
# OMEGA Trial example as presented in          #
# Chap. 19, Sect. 19.1.2.                      #
#----------------------------------------------#

optimal_bound(1.85625,250,0.0147,
              0.3,0.2,0.2,0.025,1.3,287)
```

```
[illegible]
print(Freq, [illegible], print.gap [illegible])
[illegible]
print(cbind(n_gaps_[illegible]))
print(round(n_dcon_ff[illegible]))

}

#****************************************************#
# Exemplary Calls                                    #
#****************************************************#

######################################################
# Specify                                            #
# [illegible] lsfotime,                              #
# beta_ls, p1_wrong, sta_scop,                       #
# d_li_scop                                          #
######################################################

#____________________________________________________#
# OMEGA trial example as presented in                #
# Chap. 19, Sect. 19.1.2                             #
#____________________________________________________#

calinal_bound(1.8[illegible],0.[illegible]7,
  [illegible]
```

Part III
Confirmatory Multiple Test Problem

As discussed in Parts I and II of this book, there are clinical situations where a single primary endpoint, either a composite or a particularly relevant component, is sufficient to reflect the efficacy of a new treatment. Exemplary clinical situations have been outlined in Chap. 4. A single relevant component is the endpoint of choice if enough events are expected to yield a predefined power. A single composite endpoint is suitable, if a component alone does not yield enough events and if all components are of similar clinical relevance. However, there also exist situations where a composite endpoint defines the only option to guarantee a certain power level, but the components of interest are of different severity. Such a composite endpoint is difficult to interpret if its effect is to be judged separately from the effects of the individual components. Moreover, there also exist situations where it is not clear in the planning stage if a single component provides enough events, so that a composite endpoint might be considered as another option. In these situations, the efficacy claim of the trial should be based on a multiple confirmatory test problem assessing both the composite and the most relevant component(s). Thereby, the challenge is to improve interpretation without too much loss in power compared to the single-endpoint problem assessing exclusively the composite endpoint. This can be achieved by several options when defining the corresponding multiple test problem. These options reflect different study goals which correspond to varying strengths of the efficacy claims. A "strong" claim usually provides more evidence and corresponds to a higher information level; on the other hand, this also comes along with a higher loss in power.

In this part of the book, we therefore introduce the underlying multiple test problem for a clinical trial considering a composite endpoint and a single component as primary efficacy measures. Thus, the focus lays on a confirmatory test problem for several endpoints. It is important to differentiate between the formal efficacy claim defined by the (multiple) test problem the trial is powered for and other test hypotheses that are additionally assessed but where a rejection of the corresponding null hypotheses is not obligatory to conclude efficacy of the new intervention. For these latter confirmatory analyses, the trial usually does not provide a predefined power. These claims can be seen as "nice to have" but not as necessary to demonstrate efficacy. In the following, the focus lies on the formulation of the confirmatory efficacy claim. Strategies for an additional confirmatory testing of less relevant components or for a descriptive analysis are provided in Part V of this book. Note that all multiple test problems considered in here are based on local test hypotheses and corresponding tests for the individual endpoints. These local test hypotheses and tests are given as defined in Chaps. 5 and 6 in Part II of this book. Therefore, in this part of the book, we focus on the general formulation of multiple test problems without repeating the definitions for the local tests. For the ease of presentation, it will be assumed in the following that there exists one particularly harmful and relevant component which corresponds to EP_1 without loss of generality. The multiple test problem should thus assess the composite endpoint and EP_1. However, the principles presented in the following might also be applied when there are several main components to be included in the formal efficacy claim.

Chapter 8
Correlation Between Test Statistics

It is evident that the composite endpoint is, by definition, correlated to its components as every event related to a component is also an event in the composite. This correlation might be incorporated in different ways in the planning stage and/or the analysis of the trial and is hence of general interest. Therefore, in this chapter the correlations between a composite and a single component and the correlation between two individual components are deduced for binary and time-to-event endpoints.

8.1 Composite Binary Endpoints

For a composite binary endpoint and its components, the correlations between the test statistics are given as

$$Corr\left(T_{CE}, T_{EP_l,\delta_l}\right) = \frac{\sigma_{CE,\Delta_{CE}}}{\sigma_{CE,0}} \cdot \frac{\sigma_{EP_l,\Delta_l}}{\sigma_{EP_l,\delta_l}} \cdot Cov\left(T_{CE}, T_{EP_l,\delta_l}\right), \tag{8.1}$$

$$Corr\left(T_{EP_l,\delta_l}, T_{EP_s,\delta_s}\right) = \frac{\sigma_{EP_l,\Delta_l}}{\sigma_{EP_l,\delta_l}} \cdot \frac{\sigma_{EP_s,\Delta_s}}{\sigma_{EP_s,\delta_s}} \cdot Cov\left(T_{EP_l,\delta_l}, T_{EP_s,\delta_s}\right), \ l, s = 1, \ldots, k. \tag{8.2}$$

Under the corresponding null hypotheses (5.1) and (5.2), respectively, it thus holds

$$Corr\left(T_{CE}, T_{EP_l,\delta_l}\right) = Cov\left(T_{CE}, T_{EP_l,\delta_l}\right),$$
$$Corr\left(T_{EP_l,\delta_l}, T_{EP_s,\delta_s}\right) = Cov\left(T_{EP_l,\delta_l}, T_{EP_s,\delta_s}\right), \ l, s = 1, \ldots, k.$$

G. Rauch et al., *Planning and Analyzing Clinical Trials with Composite Endpoints*, Springer Series in Pharmaceutical Statistics,
https://doi.org/10.1007/978-3-319-73770-6_8

The covariances are approximated by

$$
\begin{aligned}
Cov\left(T_{CE}, T_{EP_l,\delta_l}\right) &\approx Cov\left(\frac{T_{CE}\cdot\widehat{\sigma}_{CE,0}}{\sigma_{CE,0}}, \frac{T_{EP_l,\delta_l}\cdot\widehat{\sigma}_{EP_l,\delta_l}}{\sigma_{EP_l,\delta_l}}\right)\\
&= \frac{Cov\left(\widehat{p}^I_{CE}-\widehat{p}^C_{CE}, \widehat{p}^I_{EP_l}-\widehat{p}^C_{EP_l}\right)}{\sigma_{CE,0}\cdot\sigma_{EP_l,\delta_l}}\\
&= \frac{Corr\left(X^C_{CE}, X^C_{EP_l}\right)\frac{\sqrt{p^C_{CE}(1-p^C_{CE})p^C_{EP_l}\left(1-p^C_{EP_l}\right)}}{n_C}}{\sigma_{CE,0}\cdot\sigma_{EP_l,\delta_l}}\\
&+\frac{Corr\left(X^I_{CE}, X^I_{EP_l}\right)\frac{\sqrt{p^I_{CE}(1-p^I_{CE})p^I_{EP_l}\left(1-p^I_{EP_l}\right)}}{n_I}}{\sigma_{CE,0}\cdot\sigma_{EP_l,\delta_l}},
\end{aligned}
$$

and

$$
\begin{aligned}
Cov\left(T_{EP_l,\delta_l}, T_{EP_s,\delta_s}\right) &\approx Cov\left(\frac{T_{EP_l,\delta_l}\cdot\widehat{\sigma}_{EP_l,\delta_l}}{\sigma_{EP_l,\delta_l}}, \frac{T_{EP_s,\delta_s}\cdot\widehat{\sigma}_{EP_s,\delta_s}}{\sigma_{EP_s,\delta_s}}\right)\\
&= \frac{Cov\left(\widehat{p}^I_{EP_l}-\widehat{p}^C_{EP_l}, \widehat{p}^I_{EP_s}-\widehat{p}^C_{EP_s}\right)}{\sigma_{EP_l,\delta_l}\cdot\sigma_{EP_s,\delta_s}}\\
&= \frac{Corr\left(X^C_{EP_l}, X^C_{EP_s}\right)\frac{\sqrt{p^C_{EP_l}\left(1-p^C_{EP_l}\right)p^C_{EP_s}\left(1-p^C_{EP_s}\right)}}{n_C}}{\sigma_{EP_l,\delta_l}\cdot\sigma_{EP_s,\delta_s}}\\
&+\frac{Corr\left(X^I_{EP_l}, X^I_{EP_s}\right)\frac{\sqrt{p^I_{EP_s}(1-p^I_{EP_s})p^I_{EP_l}\left(1-p^I_{EP_l}\right)}}{n_I}}{\sigma_{EP_l,\delta_l}\cdot\sigma_{EP_s,\delta_s}},
\end{aligned}
$$

where

$$
Corr\left(X^I_{CE}, X^I_{EP_l}\right) = \frac{p^I_{EP_l}\left(1-p^I_{CE}\right)}{\sqrt{p^I_{EP_l}\left(1-p^I_{EP_l}\right)p^I_{CE}\left(1-p^I_{CE}\right)}},
$$

$$
Corr\left(X^I_{EP_l}, X^I_{EP_s}\right) = \frac{\left(\sum_{\omega\in\Omega^I_{l,s}}\omega\right)\cdot\left(1-p^I_{CE}\right)-p^I_{EP_{l,ex}}\cdot p^I_{EP_{s,ex}}}{\sqrt{p^I_{EP_l}\left(1-p^I_{EP_l}\right)p^I_{EP_s}\left(1-p^I_{EP_s}\right)}},\ l,s=1,\ldots,k,
$$

$$Corr\left(X_{CE}^{C}, X_{EP_l}^{C}\right) = \frac{p_{EP_l}^{C}\left(1-p_{CE}^{C}\right)}{\sqrt{p_{EP_l}^{C}\left(1-p_{EP_l}^{C}\right)p_{CE}^{C}\left(1-p_{CE}^{C}\right)}},$$

$$Corr\left(X_{EP_l}^{C}, X_{EP_s}^{C}\right) = \frac{\left(\sum_{\omega\in\Omega_{l,s}^{C}}\omega\right)\cdot\left(1-p_{CE}^{C}\right)-p_{EP_{l,ex}}^{C}\cdot p_{EP_{s,ex}}^{C}}{\sqrt{p_{EP_l}^{C}\left(1-p_{EP_l}^{C}\right)p_{EP_s}^{C}\left(1-p_{EP_s}^{C}\right)}},\ l,s=1,\ldots,k,$$

and

$$\Omega_{l,s}^{I} := \Omega^{I}\cap\left\{p_{EP_l}^{I}\right\}\cap\left\{p_{EP_s}^{I}\right\},\ \Omega_{l}^{C} := \Omega^{C}\cap\left\{p_{EP_l}^{C}\right\}\cap\left\{p_{EP_s}^{C}\right\}$$

denote the sets of disjoint overlap probabilities for the two groups containing $p_{EP_l}^{I}$ and $p_{EP_s}^{I}$ or $p_{EP_l}^{C}$ and $p_{EP_s}^{C}$, respectively; compare Chap. 5. The above considerations are also discussed by Rauch and Kieser (2013, 2012) and Sozu et al. (2010).

8.2 Composite Time-To-First-Event Endpoints

As discussed above for binary endpoints, the correlation between the different endpoints might be incorporated in different ways in the planning stage and/or in the analysis of the trial. Note that the construction of the multistate model defined in Sect. 5.2 does not imply any assumption on statistical dependencies between the components. As a consequence, the individual cause-specific hazards can be chosen freely without any restriction. A further consequence is that the test statistics comparing the cause-specific hazards between groups are asymptotically independent (Kay and Schumacher 1982). For $l, s = 1,\ldots,k$, the correlations are thus approximately given as

$$Corr\left(W_{EP_l,\delta_l}, W_{EP_s,\delta_s}\right) = Cov\left(W_{EP_l,\delta_l}, W_{EP_s,\delta_s}\right) = 0,\ l,s=1,\ldots,k.$$

In contrast, the test statistic for the composite endpoint $W_{CE}(0)$ is clearly correlated to the test statistic of its components $W_{EP_l,\delta_l},\ l=1,\ldots,k$, as every event in a single component also corresponds to an event in the composite endpoint. The correlation depends on the corresponding hazards $\lambda_{CE}^{I}, \lambda_{CE}^{C}, \lambda_{EP_l}^{I}, \lambda_{EP_l}^{C},\ l=1,\ldots,k$, on the underlying event time distribution, and on the censoring distribution. Due to this high number of influencing parameters, a general formula for the underlying correlation is difficult to deduce. Rauch and Kieser (2012) provide simulations that demonstrate how sensitive the correlation between test statistics depends on the underlying hazards even when neglecting the influence of different event time and censoring distributions. As a consequence, it is not recommended to use methods that are based on fixed planning assumptions for the correlation between the composite and its components as correlations are likely to be heavily misspecified.

An alternative might be to plug in estimates of the correlation from (interim) study data, e.g., by using bootstrapping algorithms (Rauch and Kieser 2012).

References

Kay, R., & Schumacher, M. (1982). Unbiased assessment of treatment effects on disease recurrence and survival in clinical trials. *Statistics in Medicine, 2*, 127–161.

Rauch, G., & Kieser, M. (2012). Multiplicity adjustment for composite binary endpoints. *Methods of Information in Medicine, 51*, 309–317.

Rauch, G., & Kieser, M. (2013). An expected power approach for the assessment of composite endpoints and their components. *Computational Statistics & Data Analysis, 60*, 111–122.

Sozu, T., Sugimoto, T., & Hamasaki, T. (2010). Sample size determination in clinical trials with multiple co-primary binary endpoints. *Statistics in Medicine, 29*, 2169–2179.

Chapter 9
The Single-Stage Design

In this chapter, several approaches are presented to define a multiple confirmatory test problem for a composite endpoint and its most relevant component. Thereby, we begin by considering the standard single-stage design without interim analyses.

9.1 Formulation of the Multiple Test Problem

As motivated in Part I of this book, interpretation problems of composite endpoints might be overcome by additionally assessing the most important component within an adequate multiple test problem. This can be achieved in several ways. A number of approaches are presented below. Thereby, the individual test hypotheses and test statistics for binary and time-to-event endpoints are defined as introduced in Part I.

9.1.1 Intersection-Union Test

The most simple idea to ease interpretation of a clinical trial with a composite endpoint would be to introduce an additional test for the main component to assess if the intervention also positively influences the most important component EP_1. An intuitive multiple test strategy could thus be given by an intersection-union test (IUT). The test problem for an intersection-union test is formulated as follows

$$H_0^{IUT} : H_0^{CE} \cup H_0^{EP_1,\delta_1} \text{ versus } H_1^{IUT} : H_1^{CE} \cap H_1^{EP_1,\delta_1},\ \delta_1 \leq 0, \tag{9.1}$$

where the individual test hypotheses H_0^{CE} and $H_0^{EP_1,\delta_1}$ are defined in Part II of this book. Note that the number of events in an individual component is usually limited as this was one of the main motivations to use a composite endpoint instead of a

G. Rauch et al., *Planning and Analyzing Clinical Trials with Composite Endpoints*, Springer Series in Pharmaceutical Statistics,
https://doi.org/10.1007/978-3-319-73770-6_9

single component. Thus, the power for the intersection-union test combining two superiority test hypotheses (for the composite endpoint and for the most important component) would usually be far too low. A more realistic approach would be to formulate a non-inferiority test for the main component within the intersection-union approach, thus setting $\delta_1 < 0$. This approach has been discussed by Huque et al. (2011) and Röhmel et al. (2006). By the closed test principle, a null hypothesis is rejected at a predefined level α when all intersection null hypotheses are rejected by an appropriate test, each at the full level α. In here, this means that both individual null hypotheses H_0^{CE} and $H_0^{EP_1,\delta_1}$ must be rejected by an appropriate level-α test in order to reject H_0^{IUT}. Thus, the intersection-union test does not require adjustment of the local significance levels (Berger 1982).

9.1.1.1 Sample Size Calculation for the IUT

The sample size for the intersection-union test formulated in (9.1) can be derived by determining the smallest value of the sample size fulfilling

$$P_{H_1^{IUT}}\left(Z_{CE} \geq z_{1-\alpha} \wedge Z_{EP_1,\delta_1} \geq z_{1-\alpha}\right) \geq 1-\beta. \tag{9.2}$$

Thereby, as before, $Z_{CE}, Z_{EP_1,\delta_1}$ are used as general notations for normally distributed test statistics which are usually given by either the logrank test statistics $W_{CE}, W_{EP_1,\delta_1}$ for time-to-event endpoints or by $T_{CE}, T_{EP_1,\delta_1}$ in the binary case.

The sample size can then be derived iteratively using the following algorithm.

1. Specify the corresponding effects under H_1^{CE} and $H_1^{EP_1,\delta_1}$ to be detected and deduce the corresponding correlation ρ under the prespecified alternative H_1^{IUT} : $H_1^{CE} \cup H_1^{EP_1,\delta_1}$. According to Chap. 8, for binary endpoints the correlation is of the form

$$\rho := Corr\left(T_{CE}, T_{EP_1,\delta_1}\right) = \frac{\sigma_{CE,\Delta_{CE}}}{\sigma_{CE,0}} \cdot \frac{\sigma_{EP_1,\Delta_1}}{\sigma_{EP_1,\delta_1}} \cdot Cov\left(T_{CE}, T_{EP_1,\delta_1}\right),$$

   where the corresponding covariance between the test statistics is also provided in Chap. 8. The null variances $\sigma_{CE,0}, \sigma_{EP_l,\delta_l}$ can be approximated by plugging in the parameter assumptions under the alternative in formulas (5.5) and (5.6), respectively. For time-to-event endpoints, there exist no closed formula for the correlation. Considerations on the derivation of a valid correlation estimate are provided in Chap. 8.
2. Choose an appropriate starting value for the total sample size, e.g., $n = \max\{n_{CE}, n_{EP_1}\}$, where n_{CE}, n_{EP_1} are the required sample sizes for the corresponding single endpoint test problems as given in Sects. 5.1.3 and 5.2.3 of Part II calculated for an adjusted power of $1-\beta/2$ and local significance levels $\alpha^{CE} = \alpha^{EP_1} = \alpha$.

3. Calculate $P_{H_1^{IUT}}(Z_{CE} \geq z_{1-\alpha} \wedge Z_{EP_1,\delta_1} \geq z_{1-\alpha})$ using the corresponding bivariate normal distribution with expectation

$$\mu = \left(\frac{\Delta_{CE}}{\sigma_{CE,0}}, \frac{\Delta_1 - \delta_1}{\sigma_{EP_1,\delta_1}} \right)$$

and a correlation matrix with elements 1 at the diagonal and elements ρ at the off-diagonal. If $P_{H_1^{IUT}}(Z_{CE} \geq z_{1-\alpha} \wedge Z_{EP_1,\delta_1} \geq z_{1-\alpha}) > 1 - \beta$, set $n = n - 1$ and repeat step 3, otherwise stop.

A corresponding R code of the above algorithm is provided in Chap. 11. If the correlation is not available in the planning stage, a simple approach is to choose the sample size as the maximum of the sample sizes required for the single endpoint tests $n = \max\{n_{CE}, n_{EP_1}\}$, where n_{CE}, n_{EP_1} are the required sample sizes for the corresponding single endpoint test problems as given in Sects. 5.1.3 and 5.2.3 of Part II calculated for local significance levels $\alpha^{CE} = \alpha^{EP_1} = \alpha$ and an adjusted power of $1 - \beta/2$.

Note that the power for an intersection-union test is usually much lower than the power for the test of each individual endpoint, so that a high sample size is required to maintain a certain power level. It is intuitively clear that the required sample size depends on the choice of δ_1. Large absolute values $|\delta_1|$ correspond to local null hypotheses $H_0^{EP_1,\delta_1}$ which are easier to reject, whereas for δ_1 near 0 the corresponding efficacy claim is much stronger. Considerations on the choice of δ_1 should therefore be guided by carefully considering the medical background (What treatment effect can be expected? How relevant and harmful is an event in EP_1?) as well as by feasibility considerations. An exemplary application of the intersection-union-test principle in a real clinical trial scenario is provided in Part V of this book.

9.1.2 *Union-Intersection Test*

In a situation where it is uncertain in the planning stage if a single binary or time-to-event outcome shows enough events to guarantee a certain power, a possible approach could be to start the trial with two "candidate" primary endpoint and to base the efficacy claim of the trial on the significance of at least one of them. The two candidate endpoints can then be given by a single outcome measure and by a composite endpoint containing this variable as one component. If the component alone shows a significant and relevant treatment effect, this is sufficient to demonstrate efficacy of the new treatment, and the result is easy to interpret. A significant effect in the composite endpoint is only considered in the case that the observed treatment effect for the main component is not significant. Interpretation of the result is more difficult in this case, as several components contribute to the observed effect, but it might still be considered a better option to base the proof of efficacy on a composite endpoint than ending up with an underpowered trial. The CAPRICORN Trial (The CAPRICORN Investigators 2001) presented in Chap. 4 in Part I of this book is an illustrative example showing that the effect for a single component is not always smaller than the effect of the composite as originally intended. The strategy is thus to accept the disadvantages of a composite endpoint only in case that a composite really provides a power advantage. The corresponding test problem could then be formulated in terms of a union-intersection test (UIT).

The hypotheses of an union-intersection test are given as follows

$$H_0^{UIT} : H_0^{CE} \cap H_0^{EP_1,\delta_1} \text{ versus } H_1^{UIT} : H_1^{CE} \cup H_1^{EP_1,\delta_1},\ \delta_1 = 0, \tag{9.3}$$

where the individual null hypotheses H_0^{CE} and $H_0^{EP_1,\delta_1}$ are defined in Part II of this book. In the situation where two candidate endpoints are under consideration, a superiority test for the main component should usually be chosen ($\delta_1 = 0$) whenever a proof of efficacy for a new treatment based on a significant non-inferiority test is not sufficient.

The null hypothesis H_0^{UIT} is rejected at a predefined level α when at least one of the individual null hypotheses H_0^{CE} or $H_0^{EP_1,\delta_1}$ is rejected by an appropriate test. The test statistics for the individual null hypotheses have been defined in Part I of this book. In an union-intersection test, the individual tests for the local test hypotheses must be performed at adjusted local significance levels $\alpha^{CE}, \alpha^{EP_1} < \alpha$ to maintain the global type I error α. An intuitive approach is to define fixed adjusted local levels α^{CE} and α^{EP_1} which are chosen such that for given test statistics Z_{CE} and Z_{EP_1,δ_1} and a one-sided global level α the following inequality is fulfilled

$$P_{H_0^{UIT}}\left(Z_{CE} \geq z_{1-\alpha^{CE}} \vee Z_{EP_1,\delta_1} \geq z_{1-\alpha^{EP_1}}\right) \leq \alpha. \tag{9.4}$$

The simplest way to define such local levels is to use the standard Bonferroni ($\alpha^{CE} = \alpha^{EP_1}$) or weighted Bonferroni adjustment ($\alpha^{CE} \neq \alpha^{EP_1}$), where the local levels are chosen such that

$$\alpha^{CE} + \alpha^{EP_1} = \alpha. \tag{9.5}$$

If the correlation of the test statistics under H_0^{UIT} is known in the planning stage, the joint distribution of the test statistics can be used to define improved local significance levels with

$$\alpha^{CE} + \alpha^{EP_1} > \alpha. \tag{9.6}$$

For binary endpoints, the null correlation can be deduced as described in Chap. 8. This approach was proposed by Rauch and Kieser (2012) and is implemented in Chap. 11. Alternatively, the null correlation might be bootstrapped from the data of the control arm. This idea is applicable to binary and to time-to-event endpoints and was realized in an R-function `bootStrap_Corr()` in Chap. 11 provided in the context of the "effect consistency approach" presented below. The bootstrap approach is appealing on the one hand as no additional planning assumptions are required. On the other hand, as the correlation adjusted local levels depend on the final data, the adjusted levels can only provide a gain in power and cannot be applied to reduce the sample size in the planning stage. A corresponding R code for correlation-adjusted local levels is also given in Chap. 11.

More advanced adjustment procedures, so-called sequentially rejective approaches, allow to recover local levels of already rejected hypotheses, thereby minimizing the power loss due to multiplicity. Whereas sequentially rejective procedures should be preferred in the analysis stage, sample size calculation is much easier for fixed local levels. A simple example for a sequentially rejective procedure is the well-known Bonferroni-Holm method (Holm 1979) which always provides a power advantage compared to the Bonferroni approach. Even when basing the sample size calculation on Bonferroni-adjusted levels, the Bonferroni-Holm approach can be used and is recommended for the analysis. As the required sample size for the Bonferroni-Holm approach is always smaller than for the Bonferroni method, this defines a conservative procedure.

9.1.2.1 Sample Size Calculation for the UIT

The sample size for the union-intersection test formulated in (9.3) can be obtained by determining the smallest value of the sample size which fulfills

$$P_{H_1^{UIT}}\left(Z_{CE} \geq z_{1-\alpha^{CE}} \vee Z_{EP_1,\delta_1} \geq z_{1-\alpha^{EP_1}}\right) \geq 1-\beta. \tag{9.7}$$

Similarly as above, the sample size can then be derived iteratively using the following algorithm.

1. Specify the corresponding effects under H_1^{CE} and $H_1^{EP_1,\delta_1}$ to be detected and deduce the corresponding correlation ρ under the prespecified alternative $H_1^{UIT}: H_1^{CE} \cap H_1^{EP_1,\delta_1}$. According to Chap. 8, for binary endpoints, the correlation is of the form

$$\rho := Corr\left(T_{CE}, T_{EP_1,\delta_1}\right) = \frac{\sigma_{CE,\Delta_{CE}}}{\sigma_{CE,0}} \cdot \frac{\sigma_{EP_1,\Delta_1}}{\sigma_{EP_1,\delta_1}} \cdot Cov\left(T_{CE}, T_{EP_l,\delta_l}\right),$$

   where the corresponding covariance between the test statistics is also provided in Chap. 8. The null variances $\sigma_{CE,0}, \sigma_{EP_1,\delta_1}$ can be approximated by plugging in the parameter assumptions under the alternative in formulas (5.5) and (5.6), respectively. As mentioned before, for time-to-event endpoints, there exist no closed formula for the correlation. Considerations on the derivation of a valid correlation estimate are provided in Chap. 8.
2. Choose an appropriate starting value for the total sample size, e.g., $n = \max\{n_{CE}, n_{EP_1}\}$, where n_{CE}, n_{EP_1} are the required sample sizes for the corresponding single endpoint test problems as given in Sects. 5.1.3 and 5.2.3

of Part II calculated for a power of $1-\beta$ and adjusted local significance levels $\alpha^{CE}, \alpha^{EP_1}$ fulfilling Eq. (9.4).

3. Calculate $P_{H_1^{UIT}}(Z_{CE} \geq z_{1-\alpha} \vee Z_{EP_1,\delta_1} \geq z_{1-\alpha})$ using the corresponding bivariate normal distribution with expectation

$$\mu = \left(\frac{\Delta_{CE}}{\sigma_{CE,0}}, \frac{\Delta_1 - \delta_1}{\sigma_{EP_1,\delta_1}}\right)$$

and a correlation matrix with elements 1 at the diagonal and elements ρ at the off-diagonal. If $P_{H_1^{UIT}}(Z_{CE} \geq z_{1-\alpha} \vee Z_{EP_1,\delta_1} \geq z_{1-\alpha}) > 1-\beta$, set $n = n-1$ and repeat step 3, otherwise stop.

An R code of the above algorithm is given in Chap. 11. An ad hoc solution ignoring the correlation of the test statistics is to choose the sample size as the maximum of the corresponding sample sizes for the single endpoint test problems calculated for a power of $1-\beta$ and for prespecified adjusted significance levels $\alpha^{CE}, \alpha^{EP_1}$ fulfilling Eq. (9.4).

The power for a union-intersection test is lower than the power for the test of each individual endpoint due to the multiplicity adjustment. However, this power loss is limited, especially when the correlation under the null hypothesis H_0^{UIT} is taken into account when defining the local adjusted levels α^{CE} and α^{EP_1}. Note that the union-intersection approach does not necessarily yield confirmatory evidence for both endpoints, and thus the gain in information is smaller than for the intersection-union approach.

9.1.3 *Effect Consistency Approach*

An alternative strategy to improve the interpretation of a composite endpoint might be to introduce a consistency criterion connecting the effect of the composite with the effect of the most relevant component instead of a formal hypothesis test for the main component. Such a consistency claim could be, for example, that the component effect should not point in opposite direction to the composite effect, meaning that the one-sided p-value of the superiority test for the main component, denoted by p_{EP_1}, should not be larger than 0.5 (or equivalently that the normally distributed test statistic $Z_{EP_1,0}$ should not be negative). Alosh and Huque have proposed several such "effect consistency approaches" (ECA) (Huque and Alosh 2012; Huque et al. 2011; Alosh and Huque 2009). These ideas can be used and extended to address the particular problem of a composite endpoint and one main component as shown by Rauch et al. (2014). The advantage of this approach is that a consistency claim instead of a formal hypothesis test for EP_1 might be easier to show, and therefore the loss in power is expected to be smaller than for the intersection-union approach. The consistency-adjusted alpha allocation approach is an extension of the well-known Bonferroni-Holm method. The test strategy requires

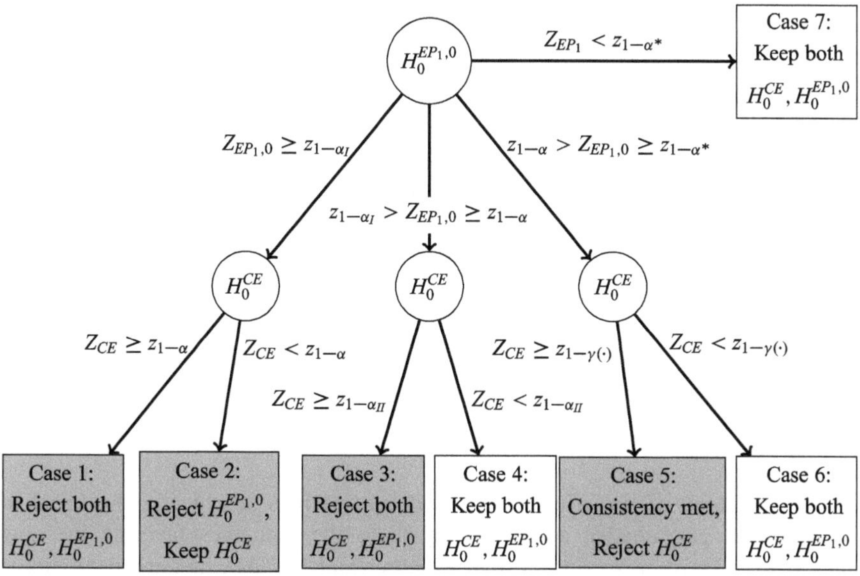

Fig. 9.1 Effect consistency approach for a composite endpoint *CE* and one main component EP_1

prespecification of the parameters $\alpha, \alpha_I, \alpha_{II}, \alpha^*, \rho$ which will be introduced in the following general description of the test procedure, compare also Fig. 9.1:

1. For a global one-sided significance level α, choose a predefined $\alpha_I < \alpha$. If $Z_{EP_1,0} \geq z_{1-\alpha_I}$, then continue to test the null hypothesis H_0^{CE} at full one-sided level α. In case the corresponding p-value p_{CE} fulfills $Z_{CE} \geq z_{1-\alpha}$ (Case 1, Fig. 9.1), then both null hypotheses $H_0^{EP_1,0}$ and H_0^{CE} can be rejected, where $H_0^{EP_1,0}$ denotes the corresponding null hypothesis of the superiority test problem. Otherwise, if $Z_{CE} < z_{1-\alpha}$ (Case 2, Fig. 9.1), only $H_0^{EP_1,0}$ is rejected. This step corresponds to simple sequentially rejective approach as described, for example, by Bretz et al. (2009).
2. If $z_{1-\alpha_I} > Z_{EP_1,0} \geq z_{1-\alpha}$, then H_0^{CE} is tested at a predefined level $\alpha_{II} < \alpha$. If $Z_{CE} \geq z_{1-\alpha_{II}}$ (Case 3, Fig. 9.1), then both null hypotheses $H_0^{EP_1,0}$ and H_0^{CE} can be rejected. In case that $Z_{CE} < z_{1-\alpha_{II}}$ (Case 4, Fig. 9.1), none of the two null hypotheses is rejected. Note that for $\alpha_{II} = \alpha_I$, steps 1 and 2 correspond to the standard (weighted) Bonferroni-Holm approach (Holm 1979).
3. The new aspect of the proposed method is that in case $Z_{EP_1,0} < z_{1-\alpha}$ we do not have to stop testing but consider a predefined consistency bound $\alpha^* > \alpha$. In case that $z_{1-\alpha} > Z_{EP_1,0} \geq z_{1-\alpha^*}$, the null hypothesis H_0^{CE} can be tested at a level $\gamma(\cdot)$, where γ is a function of the predefined parameters $\alpha, \alpha_I, \alpha_{II}, \alpha^*$ and the correlation between the test statistics ρ. If $Z_{CE} \geq z_{1-\gamma}$ (Case 5, Fig. 9.1), then H_0^{CE} can be rejected. The null hypothesis $H_0^{EP_1,0}$ cannot be rejected in this case, but the conclusion can be drawn that the effect observed for the main component

is consistent with the effect of the composite endpoint. In case $Z_{CE} < z_{1-\gamma}$ (Case 6, Fig. 9.1), neither H_0^{CE} nor $H_0^{EP_1,0}$ can be rejected.

The level $\gamma(\alpha, \alpha_I, \alpha_{II}, \alpha^*, \rho)$ can be derived by finding the value α_x which fulfills the following equation

$$\begin{aligned}\alpha - \alpha_I = &P_{H_0^{EP_1,0} \cap H_0^{CE}}(z_{1-\alpha_I} > Z_{EP_1,0} \geq z_{1-\alpha}, Z_{CE} \geq z_{1-\alpha_{II}}) \\ &+ P_{H_0^{EP_1,0} \cap H_0^{CE}}(z_{1-\alpha} > Z_{EP_1,0} \geq z_{1-\alpha^*}, Z_{CE} \geq z_{1-\alpha_x}) \qquad (9.8)\end{aligned}$$

for a known correlation between the test statistics under the assumption that both null hypotheses hold true, compare Huque et al. (2011). Then $\gamma(\alpha, \alpha_I, \alpha_{II}, \alpha^*, \rho)$ is given as

$$\gamma(\alpha, \alpha_I, \alpha_{II}, \alpha^*, \rho) = \min\{\alpha_x, \alpha_{II}\}.$$

The test strategy of the effect consistency approach is graphically displayed in Fig. 9.1. The gray-shaded areas correspond to the outcomes favoring the new intervention. The corresponding efficacy claim which has to be met in order to judge the treatment under investigation as effective reads as follows.

Consistency Efficacy Claim 1 The intervention is superior to the control with respect to the composite *or* with respect to the main component, or the intervention is superior to the control with respect to the composite *and* the main component shows a consistent effect to the composite effect (Cases 1, 2, 3 and 5, Fig. 9.1).

Note that this efficacy claim is already met if the composite shows a significant positive effect and if simultaneously the effect for the main component points in the right direction (Case 5, Fig. 9.1). Also note that if superiority for the main component can be shown (Case 2, Fig. 9.1), this suffices to meet the above efficacy claim. This seems reasonable in many applications, as most often a composite is only applied when the estimated power for the main component alone is too low. In the unlikely case that the main component shows superiority of the new intervention, it is a clear benefit to base efficacy exclusively on this endpoint, because a single-component effect is easier to interpret than a "mixed net effect" related to the composite. However, there exist other clinical trial applications where superiority for the main component alone may not suffice and some evidence for the remaining components must be guaranteed. In these situations, the above efficacy claim could be easily adapted as follows.

Consistency Efficacy Claim 2 The intervention is superior to the control with respect to the composite *and* with respect to the main component, or the intervention is superior to the control with respect to the composite *and* the main component shows a consistent effect to the composite effect (Cases 1, 3 and 5, Fig. 9.1).

The application of the effect consistency approach requires implementation which might be seen as a drawback. The main task of such a program is to solve (9.8) in order to determine $\gamma(\alpha, \alpha_I, \alpha_{II}, \alpha^*, \rho)$. This requires specification of the correlation between the test statistics under the assumption that both null hypotheses $H_0^{CE}, H_0^{EP_1}$ hold true. This correlation can be either estimated based on initial parameter guesses in the planning stage or can be calculated from the observed data by a bootstrapping algorithm (Rauch and Kieser 2012). An exemplary R code for the effect consistency approach providing both options is given in Chap. 11.

9.1.3.1 Sample Size Calculation for the ECA

Sample size calculation for the effect consistency approach is not straightforward and can only be realized by simulations. As mentioned above, the effect consistency approach was implemented in R and is provided in the Chap. 11. For simulated data following a predefined distribution, this code can be used to estimate the power and thereby to approximate the required sample size.

References

Alosh, M., & Huque, M. F. (2009). A flexible strategy for testing subgroups and overall population. *Statistics in Medicine, 28*, 3–23.

Berger, R. L. (1982). Multiparameter hypothesis testing and acceptance sampling. *Technometrics, 24*, 295–300.

Bretz, F., Maurer, W., Brannath, W., & Posch, M. (2009). A graphical approach to sequentially rejective multiple test procedures. *Statistics in Medicine, 28*, 586–604.

The CAPRICORN Investigators (2001). Effect of Carvedilol on outcome after myocardial infarction in patients with left-ventricular dysfunction: The CAPRICORN randomised trial. *The Lancet, 357*, 1385–1390.

Holm, S. (1979). A simple sequentially rejective multiple test procedure. *Scandinavian Journal of Statistics, 6*, 65–70.

Huque, F., & Alosh, M. (2012). A consistency-adjusted strategy for accommodating and underpowered primary endpoint. *Journal of Biopharmaceutical Statistics, 22*, 160–179.

Huque, F., Alosh, M., & Bhore, R. (2011). Addressing multiplicity issues of a composite endpoint and its components in clinical trials. *Journal of Biopharmaceutical Statistics, 21*, 610–634.

Rauch, G., & Kieser, M. (2012). Multiplicity adjustment for composite binary endpoints. *Methods of Information in Medicine, 51*, 309–317.

Rauch, G., Wirths, M., & Kieser, M. (2014). Consistency-adjusted alpha allocation methods for a time-to-event analysis of composite endpoints. *Computational Statistics & Data Analysis, 75*, 151–161.

Röhmel, J., Gerlinger, C., Benda, N., & Läuter, J. (2006). On testing simultaneously non-inferiority in two multiple primary endpoints and superiority in at least one of them. *Biometrical journal, 48*, 916–933.

Chapter 10
Group-Sequential and Adaptive Designs

Again, several multiple confirmatory test problems which have already been investigated within the single-stage design are considered with the aim to test both the composite and the most relevant component. In contrast to the single-stage design, the group-sequential and adaptive designs proposed in here are more flexible and thereby allow to address planning uncertainties. When several endpoints are under investigation, as considered in this part of the book, planning uncertainties are of particular relevance, as correct parameter and distribution assumptions for all endpoints are required, and thus the level of uncertainty usually increases compared to a single endpoint problem.

Note that for group-sequential and adaptive designs for several endpoints, there exist two sources of multiplicity given by the repeated hypothesis tests and by the multiple endpoint tests. In the following, it will be explained how this particular challenge can be accounted for in various situations.

10.1 Intersection-Union Test

Generally, when applying the intersection-union approach to several endpoints, there is no need to adjust for multiplicity. Therefore, the standard group-sequential boundaries can be applied without further adjustment requirements. The null hypothesis of the intersection-union test can thus be rejected whenever the smaller of both stage-wise test statistics is larger or equal than the critical value corresponding to the local level.

To apply the principles of group-sequential and adaptive designs presented in Chap. 6 in Part II of this book, the stage-wise test statistics must approximately follow a standard normal distribution under the corresponding null hypothesis. Within an intersection-union test, the stage-wise test statistics are given as the minima of the stage-wise test statistics for two endpoints. The stage-wise test statistics of an intersection-union test under the null hypothesis thus follow a

G. Rauch et al., *Planning and Analyzing Clinical Trials with Composite Endpoints*, Springer Series in Pharmaceutical Statistics,
https://doi.org/10.1007/978-3-319-73770-6_10

distribution which is given as the minimum of two standard normal distributions. This null distribution is stochastically dominated by a standard normal distribution. Therefore, when applying the decision boundaries of a standard group-sequential design to the situation of an intersection-union test, this defines a conservative test procedure, and thus type I error control is guaranteed.

10.1.1 Sample Size Calculation for the IUT

In principle, sample size calculation within a group-sequential or adaptive design for an intersection-union test can be done equivalently as described in Sect. 6.5 in Part II of this book. An exact calculation of the required sample size for a predefined power will require, along with other parameters, a valid estimate of the correlation between the two endpoint test statistics which might be unrealistic in practice. A simple solution ignoring the correlation of the test statistics is to choose the sample size as the maximum of the sample sizes for the corresponding group-sequential designs formulated separately for each individual endpoint each at a global level of α (with an adequate alpha-spending function to define the local levels) for an adjusted power of $1 - \beta/2$.

10.2 Union-Intersection Test

In Sect. 9.1.2, the use of two candidate primary endpoints has been motivated for a situation where it is uncertain in the planning stage if a single outcome shows enough events to guarantee a certain power. The treatment is claimed to be effective if at least one endpoint shows a significant treatment effect which can be assessed by a union-intersection test. Note that within a union-intersection test the local significance levels corresponding to the two endpoint tests must be adjusted in order to maintain the global significance level α.

In order to define a group-sequential or adaptive study design based on the union-intersection test, let $Z^1_{EP_1}, Z^1_{EP_2}$ denote the test statistics for the two endpoints at the interim stage. Note that for the sake of an easier notation, the two endpoints are notated as EP_1 and EP_2 throughout the section. However, as before in Sect. 9.1.2, it is reasonable to define one of the two endpoints as the composite endpoint CE. The test statistics of the cumulative data at the final analysis are referred as $Z^{1+2}_{EP_1}, Z^{1+2}_{EP_2}$. As stated above, two sources of multiplicity exist which have to be accounted for the stage-wise tests on the one hand and the multiple endpoint tests on the other hand. As before, for the sake of simplicity, we restrict our considerations on a group-sequential or adaptive design with two stages and on a multiple test problem for two candidate endpoints. In order to take into account of the repeated hypothesis tests, let $\alpha_{1,\frac{\alpha}{2}}, \alpha_{1+2,\frac{\alpha}{2}}$ be the corresponding group-sequential local significance levels for stage 1 and 2, respectively, such that

$$P_{H_0^{EP_l,\delta_l}}\left(Z^1_{EP_l,\delta_l} \geq z_{1-\alpha_{1,\frac{\alpha}{2}}} \vee \left(Z^1_{EP_l,\delta_l} < z_{1-\alpha_{1,\frac{\alpha}{2}}} \wedge Z^{1+2}_{EP_l,\delta_l} \geq z_{1-\alpha_{1+2,\frac{\alpha}{2}}}\right)\right) = \frac{\alpha}{2}, l = 1, 2, \tag{10.1}$$

where $\alpha/2$ corresponds to half of the global one-sided significance level α. Note that it is also possible to assign different fractions of the full level α to the two endpoints. However, in the situation described in here where both endpoints correspond to "candidates," an equal allocation of the full level seems appropriate and will therefore be used in the following without loss of generality. Then, a Bonferroni-type test procedure within a group-sequential or adaptive design can be defined as follows.

Definition 10.2.1 (Group-Sequential Bonferroni Approach)

Stage 1: The study stops at interim if there exists an $l \in \{1, 2\}$ such that

$$Z^1_{EP_l,\delta_l} \geq z_{1-\alpha_{1,\frac{\alpha}{2}}}. \tag{10.2}$$

All null hypotheses $H_0^{EP_l,\delta_l}$, $l = 1, 2$, fulfilling (10.2) are rejected. Otherwise, the study is continued.

Stage 2: At the final analysis, $H_0^{EP_l}$ is rejected for all $l \in \{1, 2\}$ fulfilling

$$Z^{1+2}_{EP_l,\delta_l} \geq z_{1-\alpha_{1+2,\frac{\alpha}{2}}}. \tag{10.3}$$

The group-sequential Bonferroni approach can be further improved as follows. Let $\alpha_{1,\alpha}, \alpha_{1+2,\alpha}$ be the corresponding group-sequential local significance levels for stage 1 and 2, respectively, such that

$$P_{H_0^{EP_l,\delta_l}}\left(Z^1_{EP_l,\delta_l} \geq z_{1-\alpha_{1,\alpha}} \vee \left(Z^1_{EP_l,\delta_l} < z_{1-\alpha_{1,\alpha}} \wedge Z^{1+2}_{EP_l,\delta_l} \geq z_{1-\alpha_{1+2,\alpha}}\right)\right) = \alpha, l = 1, 2, \tag{10.4}$$

for the same group-sequential adjustment rule as applied in (10.1), which implies that

$$z_{1-\alpha_{1,\frac{\alpha}{2}}} > z_{1-\alpha_{1,\alpha}}, \; z_{1-\alpha_{1+2,\frac{\alpha}{2}}} > z_{1-\alpha_{1+2,\alpha}}. \tag{10.5}$$

Then, a Bonferroni-Holm-type test procedure within a group-sequential or adaptive design can be defined as follows.

Definition 10.2.2 (Group-Sequential Bonferroni-Holm Approach)

Stage 1: The study stops at interim if there exists an $l \in \{1, 2\}$ such that

$$Z^1_{EP_l,\delta_l} \geq z_{1-\alpha_{1,\frac{\alpha}{2}}} \text{ or}$$

$$\left[z_{\alpha_{1,\frac{\alpha}{2}}} \geq Z^1_{EP_l,\delta_l} > z_{\alpha_{1,\alpha}} \text{ and } Z^1_{EP_j} \geq z_{\alpha_{1,\frac{\alpha}{2}}} \right], \; l \neq j, \; l,j = 1,2. \quad (10.6)$$

All null hypotheses $H_0^{EP_l,\delta_l}$, $l = 1, 2$, fulfilling (10.6) are rejected. Otherwise, the study is continued.

Stage 2: At the final analysis, $H_0^{EP_l,\delta_l}$ is rejected for all $l \in \{1, 2\}$ fulfilling

$$Z^{1+2}_{EP_l,\delta_l} \geq z_{1-\alpha_{1+2,\frac{\alpha}{2}}} \text{ or}$$

$$\left[z_{\alpha_{1+2,\frac{\alpha}{2}}} \geq Z^{1+2}_{EP_l,\delta_l} > z_{\alpha_{1+2,\alpha}} \text{ and } Z^{1+2}_{EP_j,\delta_j} \geq z_{\alpha_{1+2,\frac{\alpha}{2}}} \right], \; l \neq j, \; l,j = 1,2. \quad (10.7)$$

Proofs that the above test strategies maintain the global type I error rate are provided, for example, by Rauch et al. (2016) and Kieser (2005).

The above test strategies can be either used within a standard group-sequential design or within an adaptive design. Application is possible for two binary endpoints as well as for two time-to-event endpoints by means of the stage-wise test statistics provided in Part II of this book.

10.2.1 Sample Size Calculation for Group-Sequential Designs

In principle, sample size calculation within a group-sequential or adaptive design for two candidate endpoints can be done equivalently as described in Sect. 6.5 in Part II of this book. An exact calculation of the required sample size for a predefined power will require, along with other parameters, a valid estimate of the correlation between the two endpoint test statistics which might be unrealistic in practice. A simple solution ignoring the correlation of the test statistics is to choose the sample size as the maximum of the sample sizes for the corresponding group-sequential designs formulated separately for each individual endpoint each at a global level of $\alpha/2$ and a power of $1 - \beta$.

10.2.2 Sample Size Recalculation Within Adaptive Designs

The application of adaptive study designs for a multiple test problem including several endpoints is particularly appealing. In this case, the sample size in a standard fixed sequence design is usually based on a low level of evidence due to the high number of parameters which determine the required sample size. Beside an adaption of sample size, adaptive designs also allow for a variety of other design modifications; compare Sect. 6.5.2. For the specific application of adaptive designs to multiple time-to-event endpoints, care has to be taken when using a short-term endpoint at interim and a long-term time-to-event endpoint at the final analysis. In this case, the observed short-term outcomes at interim are usually correlated to non-observed outcomes of the time-to-event endpoint at interim. As a consequence, the incremental test statistic may not have the required null distribution (Jenkins et al. 2011; Bauer and Posch 2004). In this book, we exclusively discuss adaptions of the sample size. In general, arbitrary sample size recalculation rules can be applied to determine the final sample size. For a time-to-event endpoint, however, the recalculation strategy must not include information of correlated surrogate endpoints, as discussed above.

The conditional power approach presented in Sect. 6.5.2 can also be applied when using two endpoints within a union-intersection test. Thereby, in the description of the following recalculation rules, the notations introduced in Chap. 6 are used. In case of two binary endpoints, a possible recalculation rule based on the conditional power is given as follows.

1. Determine the maximum of the two observed test statistics at interim

$$t^1_{\max} := \max\left\{t^1_{CE}, t^1_{EP_1}\right\}.$$

2. Determine the required total sample size for the second stage $n_{2,recalc}$ as the smallest integer fulfilling

$$n_{2,recalc} \geq n_1 \cdot \left(\frac{-z_{1-\beta} - z_{1-\alpha_{1+2,\frac{\alpha}{2}}} \cdot \frac{\sqrt{w_1^2+w_2^2}}{w_2}}{t^1_{\max}} + \frac{w_1}{w_2} \right)^2, \tag{10.8}$$

where n_1 is the first-stage total sample size. Note that Eq. (10.8) differs from Eq. (6.19) in Sect. 6.5.2 with respect to the employed quantile. Whereas in Eq. (10.8), the quantile is $z_{1-\alpha_{1+2,\frac{\alpha}{2}}}$, the corresponding quantile in Eq. (6.19) in Sect. 6.5.2 is given by $z_{1-\alpha_{1+2}}$ which equals $z_{1-\alpha_{1+2,\alpha}}$, where the latter expression is smaller than $z_{1-\alpha_{1+2,\frac{\alpha}{2}}}$; compare (10.5).

In case of two time-to-event endpoints, the recalculation rule based on the conditional power is similar.

1. Determine the maximum of the two observed test statistics at interim

$$w^1_{\max} := \max\left\{w^1_{CE}, w^1_{EP_1}\right\}.$$

2. Determine the required number of events for the second stage as the smallest integer fulfilling

$$d_{2,recalc} \geq d_1 \cdot \left(\frac{-z_{1-\beta} - z_{1-\alpha_{1+2,\frac{\alpha}{2}}} \cdot \frac{\sqrt{w_1^2+w_2^2}}{w_2}}{w_{\max}^1} + \frac{w_1}{w_2} \right)^2. \qquad (10.9)$$

To avoid confusion with the notation, remember that $w_{\max}^1$ corresponds to the maximum of the two interim test statistics, whereas w_1, w_2 denote the weights to define the test statistic at the final analysis; compare Chap. 6 in Part II of this book.

The group-sequential Bonferroni and Bonferroni-Holm approaches presented above can thus be extended as adaptive designs with sample size recalculation. An algorithm implementing the test decision at interim based on these adaptive Bonferroni and Bonferroni-Holm approaches with sample size recalculation based on the more promising endpoint is provided in Chap. 11.

References

Bauer, P., & Posch, M. (2004). Modification of the sample size and the schedule of interim analyses in survival trials based on data inspections by H. Schäfer and H.-H. Müller, Statistics in Medicine 2001; 20: 3741–3751. *Statistics in Medicine, 23*, 1333–1334.

Jenkins, M., Stone, A., & Jennison, C. (2011). An adaptive seamless phase II/III design for oncology trials with subpopulation selection using correlated survival endpoints. *Pharmaceutical Statistics, 10*, 347–356.

Kieser, M. (2005). A note on adaptively changing the hierarchy of hypotheses in clinical trials with flexible design. *Drug Information Journal, 39*, 215–222.

Rauch, G., Schüler, S., Wirths, M., Englert, S., & Kieser, M. (2016). Adaptive designs for two candidate primary time-to-event endpoints. *Statistics in Biopharmaceutical Research, 8*, 207–216.

Chapter 11
Related Software Code

This chapter provides exemplary source code written in the software R implementing different approaches outlined in Part III of this book.

11.1 Sample Size Calculation for the IUT

```
library(mvtnorm)

###################################################
# Parameters to specify                           #
###################################################
#
# gamma   = nI/nC
# pC_CE   = Event rate in the control group
#           for the composite endpoint
# pI_CE   = Event rate in the intervention group
#           for the composite endpoint
# pC_EP   = Event rate in the control group
#           for the main component
# pI_EP   = Event rate in the intervention group
#           for the main component
# delta   = Non-inferiority margin, <= 0
# alpha   = One-sided global significance level
# power   = Desired power
###################################################



###################################################
# Calculation of required sample size for         #
# the IUT for the composite and one main component#
# were the composite is tested for superiority    #
```


G. Rauch et al., *Planning and Analyzing Clinical Trials with Composite Endpoints*, Springer Series in Pharmaceutical Statistics,
https://doi.org/10.1007/978-3-319-73770-6_11

```
# and the component for non-inferiority or        #
# superiority.                                    #
###################################################

sample_size_iut=function(pI_CE,pC_CE,pI_EP,pC_EP,
                         gamma,delta,alpha,power){

  #-------------------------------------------------#
  # Deduce the correlation rho of the test         #
  # statistics under H1_CE and H1_EP               #
  #-------------------------------------------------#

  beta = 1-power


#---------------------------------------------------#
# Calculation of the restricted maximum likelihood  #
# estimator pI_far                                  #
# under the null hypothesis restriction pC-pI=delta #
# is given as the unique solutions in (-delta;1) of #
# the maximum likelihood equation: ax^3+bx^2+cx+d=0 #
#---------------------------------------------------#

  theta = 1/gamma
  s0    = -delta
  a     = 1+theta
  b     = -(1+theta+pI_EP+theta*pC_EP+s0*(theta+2))
  c     = s0^2+s0*(2*pI_EP+theta+1)+pI_EP+theta*pC_EP
  d     = -1*pI_EP*s0*(1+s0)
  # ---------------------------------------------
  v     = (b**3)/((3*a)**3)-b*c/(6*a*a)+d/(2*a)
  u     = sign(v)*sqrt( b*b/((3*a)**2)-c/(3*a))
  w     = (1/3)*(pi+acos(v/(u**3)))
  # ---------------------------------------------
  pI_EP_far = 2*u*cos(w) - b/(3*a)
  pC_EP_far = pI_EP_far-s0


  p_CE  = (pI_CE*gamma+pC_CE)/(gamma+1)


#---------------------------------------------------#
# Calculate factor                                  #
# SIGMA = (sigmaCE_Delta/sigmaCE_0)                 #
#         *(sigmaEP_Delta/sigmaEP_delta)            #
# to be multiplied with covariance of the test      #
# statistic, compare (8.1) and (8.2) in Chap. 8.    #
# sigmaCE_0 and sigmaEP_delta are approximated      #
# by plugging in  the relative frequencies          #
# in (5.5) and (5.6) in Chap. 5.                    #
#---------------------------------------------------#

A  = sqrt(((pI_CE*(1-pI_CE))+gamma*pC_CE*(1-pC_CE))
/((1+gamma)*(p_CE*(1-p_CE))))
```

```
B  = sqrt(((pI_EP*(1-pI_EP))+(gamma*pC_EP*(1-pC_EP)))
/(pI_EP_far*(1-pI_EP_far)
+pC_EP_far*(1-pC_EP_far)*gamma))
SIGMA = A*B

#---------------------------------------------------#
# Calculate covariance of the test statistic        #
#---------------------------------------------------#

  if((pC_EP*(1-pC_CE))<=0.00001){
    CorrC=0
  }else{
    CorrC=(pC_EP*(1-pC_CE))/sqrt(
      pC_EP*(1-pC_EP)*pC_CE*(1-pC_CE))
  }
  C = sqrt(pC_CE*(1-pC_CE)*pC_EP*(1-pC_EP))
  D = sqrt(p_CE*(1-p_CE)*(1+(1/gamma)))
  E = sqrt(((1/gamma)*pI_EP_far*
            (1-pI_EP_far)+pC_EP_far*(1-pC_EP_far)))

  if((pI_EP*(1-pI_CE))<=0.00001){
    CorrI=0
  }else{
    CorrI=(pI_EP*(1-pI_CE))/sqrt(
      pI_EP*(1-pI_EP)*pI_CE*(1-pI_CE))
  }

  F = sqrt(pI_CE*(1-pI_CE)*pI_EP*(1-pI_EP))
  G = sqrt((1+gamma)*p_CE*(1-p_CE))
  H = sqrt((pI_EP_far*(1-pI_EP_far)
          +pC_EP_far*(1-pC_EP_far)*gamma))
  Cov_CE_EP = CorrC*(C/(D*E))+CorrI*(F/(G*H))

#---------------------------------------------------#
# Calculate correlation matrix of the test statistic#
#---------------------------------------------------#

Corr_CE_EP = SIGMA*Cov_CE_EP

rho        = matrix(data=c(1,Corr_CE_EP,Corr_CE_EP,1)
,nrow = 2,ncol = 2)

#---------------------------------------------------#
# Calculate sample size for each endpoint           #
# separately at a power of 1-beta/2.                #
# The maximum of these sample sizes defines the     #
# starting value to calculate IUT sample size.      #
#---------------------------------------------------#

  Delta_CE  = pC_CE - pI_CE
  Delta_EP  = pC_EP - pI_EP

  nCE = (1+(1/gamma))*((qnorm(1-alpha)
        *sqrt((gamma+1)*pC_CE*(1-pC_CE))
```

```
            + qnorm(1-(beta/2))*sqrt(pI_CE*(1-pI_CE)
            +gamma*pC_CE*(1-pC_CE)))/Delta_CE)^2
  nEP = (1+(1/gamma))*((qnorm(1-alpha)
            *sqrt(((pC_EP-delta)*(1-(pC_EP-delta)))
            +(gamma*pC_EP*(1-pC_EP)))
            +qnorm(1-(beta/2))*sqrt(pI_EP*(1-pI_EP)
            +gamma*pC_EP*(1-pC_EP)))/
(Delta_EP-delta))^2

  n   = max(ceiling(nCE), ceiling(nEP))


#----------------------------------------------------#
# Iteratively calculate the sample size for the IUT #
# by using the bivariate normal distribution        #
# with correlation matrix as specified above        #
#----------------------------------------------------#

Delta_CE   = pC_CE - pI_CE
Delta_EP   = pC_EP - pI_EP
nI         = n/(1+(1/gamma))
nC         = n/(gamma+1)
Var_CE0    = p_CE*(1-p_CE)*(nI+nC)/(nI*nC)
meanCE     = (pC_CE-pI_CE)/sqrt(Var_CE0)

Var_EP_far= ((pI_EP_far*(1-pI_EP_far))
+(pC_EP_far*(1-pC_EP_far)*gamma))/nI
meanEP     = (pC_EP-pI_EP-delta)/sqrt(Var_EP_far)

pow = pmvnorm(lower=c(qnorm(1-alpha),qnorm(1-alpha))
,upper=c(Inf,Inf),mean=c(meanCE,meanEP)
,corr=rho)[1]

  while(pow>1-beta){
n=n-1

nI         = n/(1+(1/gamma))
nC         = n/(gamma+1)
Var_CE0    = p_CE*(1-p_CE)*(nI+nC)/(nI*nC)
meanCE     = (pC_CE-pI_CE)/sqrt(Var_CE0)

Var_EP_far= ((pI_EP_far*(1-pI_EP_far))
+(pC_EP_far*(1-pC_EP_far)*gamma))/nI
meanEP     = (pC_EP-pI_EP-delta)/sqrt(Var_EP_far)

pow = pmvnorm(lower=c(qnorm(1-alpha)
,qnorm(1-alpha))
,upper=c(Inf,Inf)
,mean=c(meanCE,meanEP)
,corr=rho)[1]
  }

  n =n+1
  nI         = n/(1+(1/gamma))
```

```
  nC        = n/(gamma+1)
  Var_CE0   = p_CE*(1-p_CE)*(nI+nC)/(nI*nC)
  meanCE    = (pC_CE-pI_CE)/sqrt(Var_CE0)

  Var_EP_far= ((pI_EP_far*(1-pI_EP_far))
             +(pC_EP_far*(1-pC_EP_far)*gamma))/nI
  meanEP = (pC_EP-pI_EP-delta)/sqrt(Var_EP_far)

  pow=pmvnorm(lower=c(qnorm(1-alpha),qnorm(1-alpha))
           ,upper=c(Inf,Inf),mean=c(meanCE,meanEP)
           ,corr=rho)[1]
  print("Sample size n_I, n_C and related power")
  return(c(nI,nC,pow))
}


#*************************************#
# Exemplary calls                     #
#*************************************#

sample_size_iut(0.15,0.35,0.0375
                ,0.0875,1,-0.05,0.025,0.8)
sample_size_iut(0.15,0.35,0.0375
                ,0.0875,2,-0.05,0.025,0.8)
```

11.2 Sample Size Calculation for the UIT

```
library(mvtnorm)

##################################################
# Calculation of required sample size for        #
# the UIT for the composite and one main component#
# were the composite is tested for superiority   #
# and the component for non-inferiority or       #
# superiority                                    #
##################################################

##################################################
# Parameters to specify                          #
##################################################
#
# gamma    = nI/nC
# pC_CE    = Event rate in the control group
#            for the composite endpoint
# pI_CE    = Event rate in the intervention group
#            for the composite endpoint
# pC_EP    = Event rate in the control group
#            for the main component
# pI_EP    = Event rate in the intervention group
#            for the main component
```

```
# delta    = Non-inferiority margin, <= 0
# alpha    = One-sided global significance level
# power    = Desired power
###################################################


sample_size_uit=function(pI_CE,pC_CE,pI_EP,pC_EP,
                         gamma, delta,alpha,power){

#-----------------------------------------------#
# Deduce the correlation rho of the test        #
# statistics under H1_CE and H1_EP              #
#-----------------------------------------------#

beta = 1-power


#-------------------------------------------------#
# Calculation of the restricted maximum likelihood  #
# estimator pI_far                                #
# under the null hypothesis restriction pC-pI=delta #
# is given as the unique solutions in (-delta;1) of #
# the maximum likelihood equation: ax^3+bx^2+cx+d=0 #
#-------------------------------------------------#

  theta = 1/gamma
  s0    = -delta
  a     = 1+theta
  b     = -(1+theta+pI_EP+theta*pC_EP+s0*(theta+2))
  c    = s0^2+s0*(2*pI_EP+theta+1)+pI_EP+theta*pC_EP
  d    = -1*pI_EP*s0*(1+s0)
  # -------------------------------------------
  v      = (b**3)/((3*a)**3)-b*c/(6*a*a)+d/(2*a)
  u      = sign(v)*sqrt( b*b/((3*a)**2)-c/(3*a))
  w      = (1/3)*(pi+acos(v/(u**3)))
  # -------------------------------------------
  pI_EP_far = 2*u*cos(w) - b/(3*a)
  pC_EP_far = pI_EP_far-s0


p_CE  = (pI_CE*gamma+pC_CE)/(gamma+1)

#-------------------------------------------------#
# Calculate factor                               #
# SIGMA = (sigmaCE_Delta/sigmaCE_0)              #
#         *(sigmaEP_Delta/sigmaEP_delta)         #
# to be multiplied with covariance of the test   #
# statistic, compare (8.1) and (8.2) in Chap. 8, #
# sigmaCE_0 and sigmaEP_delta are approximated   #
# by plugging in the relative frequencies        #
# in (5.5) and (5.6) in Chap. 5.                 #
#-------------------------------------------------#

A   = sqrt(((pI_CE*(1-pI_CE))+gamma*pC_CE*(1-pC_CE
```

```
              ))/((1+gamma)*(p_CE*(1-p_CE))))
B    = sqrt(((pI_EP*(1-pI_EP))+(gamma*pC_EP*(1-pC_EP
             )))/(pI_EP_far*(1-pI_EP_far)+pC_EP_far*(
           1-pC_EP_far)*gamma))
SIGMA = A*B

#---------------------------------------------------#
# Calculate covariance of the test statistic        #
#---------------------------------------------------#

if((pC_EP*(1-pC_CE))<=0.00001){
     CorrC=0
}else{
     CorrC=(pC_EP*(1-pC_CE))/sqrt(
       pC_EP*(1-pC_EP)*pC_CE*(1-pC_CE))
}

C = sqrt(pC_CE*(1-pC_CE)*pC_EP*(1-pC_EP))
D = sqrt(p_CE*(1-p_CE)*(1+(1/gamma)))
E = sqrt(((1/gamma)*pI_EP_far*
             (1-pI_EP_far)+pC_EP_far*(1-pC_EP_far)))

if((pI_EP*(1-pI_CE))<=0.00001){
     CorrI=0
}else{
     CorrI=(pI_EP*(1-pI_CE))/sqrt(
       pI_EP*(1-pI_EP)*pI_CE*(1-pI_CE))
}

F = sqrt(pI_CE*(1-pI_CE)*pI_EP*(1-pI_EP))
G = sqrt((1+gamma)*p_CE*(1-p_CE))
H = sqrt((pI_EP_far*(1-pI_EP_far)+pC_EP_far
          *(1-pC_EP_far)*gamma))

Cov_CE_EP = CorrC*(C/(D*E))+CorrI*(F/(G*H))

#---------------------------------------------------#
# Calculate correlation matrix of the test statistic#
#---------------------------------------------------#

Corr_CE_EP = SIGMA*Cov_CE_EP

rho       = matrix(data =c(1,Corr_CE_EP,Corr_CE_EP,1)
                   , nrow = 2, ncol = 2)

#---------------------------------------------------#
# Calculate sample size for each endpoint           #
# separately at a power of 1-beta and a level of    #
# alpha/2.                                          #
# The maximum of these sample sizes defines the     #
# starting value to calculate IUT sample size.      #
#---------------------------------------------------#

Delta_CE  = pC_CE - pI_CE
```

```
Delta_EP  = pC_EP - pI_EP

nCE = (1+(1/gamma))*((qnorm(1-alpha/2)*
      sqrt((gamma+1)*pC_CE*(1-pC_CE))+
      qnorm(1-(beta))*sqrt(pI_CE*(1-pI_CE)+
      gamma*pC_CE*(1-pC_CE)))/Delta_CE)^2
nEP = (1+(1/gamma))*((qnorm(1-alpha/2)*
      sqrt(((pC_EP-delta)*(1-(pC_EP-delta))+
      (gamma*pC_EP*(1-pC_EP)))+
      qnorm(1-(beta))*sqrt(pI_EP*(1-pI_EP)+
      gamma*pC_EP*(1-pC_EP)))/(Delta_EP-delta))^2

n   = max(ceiling(nCE), ceiling(nEP))


#---------------------------------------------------#
# Iteratively calculate the sample size for the IUT #
# by using the bivariate normal distribution        #
# with correlation matrix as specified above        #
#---------------------------------------------------#


Delta_CE  = pC_CE - pI_CE
Delta_EP  = pC_EP - pI_EP

nI        = n/(1+(1/gamma))
nC        = n/(gamma+1)
Var_CE0   = p_CE*(1-p_CE)*(nI+nC)/(nI*nC)
meanCE    = (pC_CE-pI_CE)/sqrt(Var_CE0)

Var_EP_far= ((pI_EP_far*(1-pI_EP_far))+(
            pC_EP_far*(1-pC_EP_far)*gamma))/nI
meanEP    = (pC_EP-pI_EP-delta)/sqrt(Var_EP_far)

pow = pmvnorm(lower=c(qnorm(1-alpha/2)
 ,qnorm(1-alpha/2))
     ,upper=c(Inf,Inf),mean=c(meanCE,meanEP)
     ,corr=rho)[1]

   while(pow>1-beta){
    n=n-1

    nI        = n/(1+(1/gamma))
    nC        = n/(gamma+1)
    Var_CE0   = p_CE*(1-p_CE)*(nI+nC)/(nI*nC)
    meanCE    = (pC_CE-pI_CE)/sqrt(Var_CE0)

    Var_EP_far= ((pI_EP_far*(1-pI_EP_far))+(
                pC_EP_far*(1-pC_EP_far)*gamma))/nI
    meanEP    = (pC_EP-pI_EP-delta)/sqrt(Var_EP_far)

    pow = 1-pmvnorm(lower=c(-Inf,-Inf)
          ,upper=c(qnorm(1-alpha/2)
                   ,qnorm(1-alpha/2))
```

```
            ,mean=c(meanCE,meanEP)
            ,corr=rho)[1]
      }


    n =n+1
    nI         = n/(1+(1/gamma))
    nC         = n/(gamma+1)
    Var_CE0    = p_CE*(1-p_CE)*(nI+nC)/(nI*nC)
    meanCE     = (pC_CE-pI_CE)/sqrt(Var_CE0)

    Var_EP_far= ((pI_EP_far*(1-pI_EP_far))+(
                pC_EP_far*(1-pC_EP_far)*gamma))/nI
    meanEP     = (pC_EP-pI_EP-delta)/sqrt(Var_EP_far)

    pow = 1-pmvnorm(lower=c(-Inf,-Inf)
                    ,upper=c(qnorm(1-alpha/2)
                             ,qnorm(1-alpha/2))
                    ,mean=c(meanCE,meanEP)
                    ,corr=rho)[1]
    return(c(n,pow))

}



#*************************************#
# Exemplary calls                     #
#*************************************#

sample_size_uit(0.15,0.35,0.0375
                ,0.0875,1,-0.05,0.025,0.8)
sample_size_uit(0.15,0.35,0.0375
                ,0.0875,2,-0.05,0.025,0.8)
```

11.3 Correlation-Adjusted Local Levels

```
library(mvtnorm)

##################################################
# Deduce the correlation rho of the test         #
# statistics under H0_CE and H0_EP               #
# for a binary composite endpoint and one main   #
# component                                      #
#------------------------------------------------#
# Alternatively, the correlation can be          #
# bootstrapped from the data to estimate the     #
# correlation (also applicable for time-to-event #
# endpoints) by means of the R-function          #
# bootStrap_corr()                               #
```

```
# provided in Section 11.D                              #
#########################################################

#########################################################
# Parameters to specify                                 #
#########################################################
#
# nI      = Sample size in the intervention group
# nC      = Sample size in the control group
# pC_CE   = Event rate in the control group
#           for the composite endpoint under H0_CE,
#           must be larger than pC_EP
#           The event rate in the intervention
#           group under H0_CE is equal to pC_CE
# pC_EP   = Event rate in the control group
#           for the main component under H0_EP.
#           The event rate in the intervention
#           group for the main component under
#           H0_EP is defined by pC_EP-pI_EP=delta
# delta   = Non-inferiority margin, <= 0
#########################################################

null_corr<-function(pC_EP,delta,pC_CE,n_I,n_C){

if(pC_CE<pC_EP){
  print("pC_CE must be larger than pC_EP")
}

gamma = n_I/n_C
pI_CE = pC_CE
pI_EP = pC_EP-delta


help_C = sqrt(pC_CE*(1-pC_CE)*pC_EP*(1-pC_EP))
help_I = sqrt(pI_CE*(1-pI_CE)*pI_EP*(1-pI_EP))
var_CE = (gamma+1)/n_I*pC_CE*(1-pC_CE)
var_EP = 1/n_I*pI_EP*(1-pI_EP)+1/n_C*pC_EP*(1-pC_EP)
Cov_C  = (help_C/n_C*pC_EP
         *(1-pC_CE)/help_C)/sqrt(var_CE*var_EP)
Cov_I  = (help_I/n_I*pC_EP
         *(1-pC_CE)/help_I)/sqrt(var_CE*var_EP)
Corr   =  Cov_C+Cov_I

print("Correlation of the test statistics")
print("under H0_CE and H0_EP")
print(Corr)
return(Corr)
}

#########################################################
# Deduce the correlation-adjusted local level           #
# to test EP1 for a fixed global level alpha and        #
# a fixed local level alpha_CE for the composite        #
#########################################################
```

```
adjust_levels<-function(alpha, alpha_CE, rho){

  alpha_EP<-alpha-alpha_CE
  correl<-diag(2)
  correl[1,2]<-rho
  correl[2,1]<-rho


  x<-1-pmvnorm(lower=c(-Inf, -Inf)
    ,upper=c(qnorm(1-alpha_CE),qnorm(1-alpha_EP))
    ,mean=c(0,0), corr=correl)[1]

  while (x<alpha){

    alpha_EP<-alpha_EP+0.000001

    x<-1-pmvnorm(lower=c(-Inf, -Inf)
    ,upper=c(qnorm(1-alpha_CE),qnorm(1-alpha_EP))
    ,mean=c(0,0), corr=correl)[1]
  }

print("Global and local levels")
print("for the composite and the main component")
print(c(alpha,alpha_CE,alpha_EP))

}

#*************************************#
# Exemplary calls                     #
#*************************************#

rho <- null_corr(0.2,-0.01,0.3,100,100)

adjust_levels(0.025, 0.02, rho)
```

11.4 Implementation of the ECA

11.4.1 ECA for Binary Endpoints

```
library(survival)
library(mvtnorm)


#####################################################
# Main function to bootstrap the correlation of the #
# test statistics under the intersection of both    #
# null hypotheses related to the two endpoints      #
#####################################################
```

```
###################################################
# Parameters to specify                           #
###################################################
#
#
# Before you start the program it is necessary to
# import your data into R.
#
# Required structure of the data set "data":
#
# 1.column (id)      = Patient ID
# 2.column (status1) = Binary event indicator for
#                      EP1 (Main Component)
# 3.column (status2) = Binary event indicator for
#                      EP2 (Composite)
# 4.column (group)   = Group allocation
#--------------------------------------------------
# Other required parameters
#--------------------------------------------------
# sampleCount = Number of bootstrap samples to be
#               drawn for correlation estimation
###################################################

bootStrap_corr<-function(data, sampleCount){
#---------------------------------------------#
# Calculate sample size of the groups         #
#---------------------------------------------#

   samplesize_C <-length(data[,1])-sum(data[,4])
   samplesize_I <-sum(data[,4])


#----------------------------------------------#
# Use the control data set for bootstrapping   #
# a hypothetical control and intervention      #
# sample under the intersection of both        #
# null hypotheses                              #
#----------------------------------------------#
    group_C_Data <- data[data[,4] == 0,]
    group_I_Data <- group_C_Data
    group_I_Data[,4] = 1


#----------------------------------------------#
# Extract "sampleCount" samples out of the     #
# given given data set of the control group    #
# where each sample has the original size      #
#----------------------------------------------#

g_I_Rows <- nrow(group_I_Data)
g_C_Rows <- nrow(group_C_Data)
tmp_I <- replicate(sampleCount,
group_I_Data[sample.int(g_I_Rows
,samplesize_I, T),],simplify=F)
```

```
tmp_C <- replicate(sampleCount,
group_C_Data[sample.int(g_C_Rows
,samplesize_C, T),],simplify=F)

sampleData<-mapply(rbind,tmp_I,tmp_C,SIMPLIFY =F)

#-----------------------------------------------#
# "getTestStat": Help function to calculate the #
# test statistic of both endpoints              #
#-----------------------------------------------#

getTestStat <-function(data,
endpoint_indicator = T){

#-----------------------------------------------#
# If endpoint_indicatpor is TRUE, than the      #
# first endpoint is used otherwise the second   #
#-----------------------------------------------#


      if(endpoint_indicator){
        datanew <- data[,c(2,4)]
      }else{
        datanew <- data[,c(3,4)]
      }

      data_C <- datanew[datanew[,2] == 0,]
      data_I <- datanew[datanew[,2] == 1,]

      p_C <- mean(data_C[,1])
      p_I <- mean(data_I[,1])
      p <- (p_C*samplesize_C+p_I*samplesize_I)/(
        samplesize_C+samplesize_C)
      var_0 <- p*(1-p)*(samplesize_C+samplesize_C)/(
        samplesize_C*samplesize_C)
      teststat<- (p_C-p_I)/sqrt(var_0)

      return(teststat)
    }

#-----------------------------------------------#
# For each bootstrap sample the test statistic  #
# for both endpoints are calculated through the #
# function "getTestStat"                        #
#-----------------------------------------------#

Teststat_ep <- mapply(getTestStat, sampleData)
Teststat_ce <- mapply(getTestStat, sampleData
,MoreArgs=list(endpoint_indicator=F))

#-----------------------------------------------#
# Calculate the correlation between the         #
# test stastistics                              #
#-----------------------------------------------#
```

```
return(cor(Teststat_ce, Teststat_ep))
}




#####################################################
# Main function to calculate the correlation        #
# dependent parameter alpha_rho, the test statistics#
# of both endpoints and the test decision of the    #
# effect consistency approach (ECA)                 #
#####################################################


###################################################
# Parameters to specify                           #
###################################################
# Before you start the program it is necessary to
# import your data into R.
#
# Required structure of the data set "data":
#
# 1.column (id)      = Patient ID
# 2.column (times)   = Event time
# 3.column (status1) = Binary event indicator for
#                      ep2 (Main Component)
# 4.column (status2) = Binary event indicator for
#                      EP2 (Composite)
# 5.column (group)   = Group allocation
#--------------------------------------------------
# Other required parameters
#--------------------------------------------------
# alpha1    = Parameter of the ECA see Chap. 9.1.3
# alpha2    = Parameter of the ECA see Chap. 9.1.3
# alpha     = Global one-sided significance levek
# alphaStern  = Consistency parameter of the ECA
#                compare Chap. 9.1.3
# rho       = Correlation between the test statistics
#             under the intersection null hypothesis
###################################################

testdecision<-function(data,alpha1,
                       alpha2,alpha,alphaStern,rho){

  #-------------------------------------------#
  # Calculate sample size of the groups       #
  #-------------------------------------------#

  samplesize_C <-length(data[,1])-sum(data[,4])
  samplesize_I <-sum(data[,4])
```

```
#------------------------------------------------#
# Calculate the test statistics of               #
# ep2 (Main Component and EP2 (Composite)        #
#------------------------------------------------#



data_ep1 <- data[,c(2,4)]
data_ep2 <- data[,c(3,4)]


data_ep1_C <- data_ep1[data_ep2[,2] == 0,]
data_ep1_I <- data_ep1[data_ep2[,2] == 1,]

data_ep2_C <- data_ep2[data_ep2[,2] == 0,]
data_ep2_I <- data_ep2[data_ep2[,2] == 1,]

p_ep1_C <- mean(data_ep1_C[,1])
p_ep1_I <- mean(data_ep1_I[,1])
p_ep1 <- (p_ep1_C*samplesize_C+
  p_ep1_I*samplesize_I)/(samplesize_C
  +samplesize_C)
var_ep1_0 <- p_ep1*(1-p_ep1)*(samplesize_C
  +samplesize_C)/(samplesize_C*samplesize_C)
teststat1<- (p_ep1_C-p_ep1_I)/sqrt(var_ep1_0)

p_ep2_C <- mean(data_ep2_C[,1])
p_ep2_I <- mean(data_ep2_I[,1])
p_ep2 <- (p_ep2_C*samplesize_C
  +p_ep2_I*samplesize_I)/(samplesize_C
  +samplesize_C)
var_ep2_0 <- p_ep2*(1-p_ep2)*(samplesize_C
  +samplesize_C)/(samplesize_C*samplesize_C)
teststat2<- (p_ep2_C-p_ep2_I)/sqrt(var_ep2_0)

#------------------------------------------------#
# Calculate alpha_rho by solving Eq.             #
# (9.8) Chap. 9 in dependence of rho             #
#------------------------------------------------#

r12<-matrix(c(1,rho,rho,1),ncol=2)
pow<-1
alphax<-0

pow1<-(pmvnorm(lower=c(-Inf,-Inf)
               ,upper=c(qnorm(1-alpha1)
                ,qnorm(1-alpha2))
               ,mean=c(0,0),corr=r12)
       -pmvnorm(lower=c(-Inf,-Inf)
                ,upper=c(qnorm(1-alpha)
                ,qnorm(1-alpha2))
                ,mean=c(0,0),corr=r12))

boarder <- alphaStern-alpha
```

```
while(pow>= boarder){
  alphax<-alphax + 0.001
  pow2<-(pmvnorm(lower=c(-Inf,-Inf)
          ,upper=c(qnorm(1-alpha)
          ,qnorm(1-alphax))
          ,mean=c(0,0),corr=r12)
         -pmvnorm(lower=c(-Inf,-Inf)
          ,upper=c(qnorm(1-alphaStern)
          ,qnorm(1-alphax))
          ,mean=c(0,0),corr=r12))
  pow<-pow1+pow2
}

while(pow>=boarder){
  alphax<-alphax-0.0001
  pow2<-(pmvnorm(lower=c(-Inf,-Inf)
          ,upper=c(qnorm(1-alpha)
          ,qnorm(1-alphax))
          ,mean=c(0,0), corr=r12)
         -pmvnorm(lower=c(-Inf,-Inf)
          ,upper=c(qnorm(1-alphaStern)
          ,qnorm(1-alphax))
          ,mean=c(0,0), corr=r12))
  pow<-pow1-pow2
}

alphax<-alphax+0.0001
alpha_rho<-min(alphax,alpha2)


#----------------------------------------------------#
# Test decision of the ECA, compare Fig. 9.1 Chap. 9 #
#----------------------------------------------------#

if(teststat1>=qnorm(1-alpha1)
   && teststat2>=qnorm(1-alpha1)){
  print("Case 1: Reject both null hypotheses")
}
  else if(teststat1>=qnorm(1-alpha1)
  && teststat2<qnorm(1-alpha1)){
 print("Case 2: Reject only the nullhypothesis")
 print("of EP1 (Main Component)")
}
else if(teststat1<qnorm(1-alpha1)
  && teststat1>=qnorm(1-alpha)
  && teststat2>=qnorm(1-alpha2)){
  print("Case 3: Reject both null hypotheses")
}
else if(teststat1<qnorm(1-alpha1)
  && teststat1>=qnorm(1-alpha)
  && teststat2<qnorm(1-alpha2)){
  print("Case 4: Keep both null hypotheses")
}
```

```
  else if(teststat1<qnorm(1-alpha)
    && teststat1>=qnorm(1-alphaStern)
    && teststat2>=qnorm(1-alpha_rho)){
    print("Case 5: Reject the nullhypothesis")
    print(" of EP2 (Composite).")
    print("The effect of EP1 (Main Component)")
    print("is consistent.")
  }
  else if(teststat1<qnorm(1-alpha)
    && teststat1>=qnorm(1-alphaStern)
    && teststat2<qnorm(1-alpha_rho)){
    print("Case 6: Keep both null hypotheses")
  }
  else if(teststat1<qnorm(1-alphaStern) ){
    print("Case 7: Keep both null hypotheses")
  }

  print("Test statistics and p-values")
  print("of EP1 (Main Component),EP2 (Composite)")
  print(c(teststat1, 1-pnorm(teststat1)
         , teststat2,1-pnorm(teststat2)))
  print("Parameter alpha_rho")
  print(alpha_rho)
}

#************************************#
# Exemplary calls                    #
#************************************#

set.seed(150)
#------------------------------------#
# Generate data where each event in  #
# the first endpoint (main component) #
# also refers to an event in the     #
# second component (composite).      #
#------------------------------------#

#------------------------------------#
# Generate event indicators for two  #
# components                         #
#------------------------------------#

statusI<-rmultinom(20, 1, c(0.1,0.2,0.7))
statusC<-rmultinom(30, 1, c(0.2,0.3,0.5))


status1<-c(statusI[1,], statusC[1,])
status2h<-c(statusI[2,], statusC[2,])

#------------------------------------#
# Composite equals 1 whenever at     #
# least one of the components is 1   #
#------------------------------------#
```

```
status2<-rep(0,50)
status2[status2h==1 | status1==1] <-1

data<-data.frame(id=c(1:50),
                 status1=status1,
                 status2=status2,
                group=c(rep(1,20),rep(0,30)))
print(data)


corr<-bootStrap_corr(data, 500)
print(corr)


testdecision(data,0.02,0.02,0.025,0.5,corr)
```

11.4.2 ECA for Time-to-Event Endpoints

```
library(survival)
library(mvtnorm)

####################################################
# Main function to bootstrap the correlation of the#
# test statistics under the intersection of both   #
# null hypotheses related to the two endpoints     #
####################################################

###################################################
# Parameters to specify                           #
###################################################
#
#
# Before you start the program it is necessary to
# import your data into R.
#
# Required structure of the data set "data":
#
# 1.column (id)      = Patient ID
# 2.column (times)   = Event time
# 3.column (status1) = Binary event indicator for
#                      EP1 (Main Component)
# 4.column (status2) = Binary event indicator for
#                      EP2 (Composite)
# 5.column (group)   = Group allocation
#-------------------------------------------------
# Other required parameters
#-------------------------------------------------
# sampleCount = Number of bootstrap samples to be
#               drawn for correlation estimation
###################################################
```

```
bootStrap_corr<-function(data, sampleCount){

#-------------------------------------------#
# Calculate sample size of the control group  #
#-------------------------------------------#

   samplesize_C <-length(data[,1])-sum(data[,5])
   samplesize_I <-sum(data[,5])

#--------------------------------------------#
# Use the control data set for bootstrapping   #
# a hypothetical control and intervention      #
# sample under the intersection of both        #
# null hypotheses                              #
#--------------------------------------------#

    group_C_Data <- data[data[,5] == 0,]
    group_I_Data <- group_C_Data
    group_I_Data[,5] = 1

#--------------------------------------------#
# Extract "sampleCount" samples out of the     #
# given given data set of the control group    #
# where each sample has the original size      #
#--------------------------------------------#

g_I_Rows <- nrow(group_I_Data)
g_C_Rows <- nrow(group_C_Data)
tmp_I <- replicate(sampleCount,
group_I_Data[sample.int(g_I_Rows
,samplesize_I, T),],simplify=F)
tmp_C <- replicate(sampleCount,
group_C_Data[sample.int(g_C_Rows
,samplesize_C, T),],simplify=F)

sampleData<-mapply(rbind,tmp_I,tmp_C,SIMPLIFY=F)


#----------------------------------------------#
# "getTestStat" : Help function to calculate the #
# test statistic of both endpoints               #
#----------------------------------------------#


getTestStat <-function(data,endpoint_indicator=T){


#--------------------------------------------#
# If endpoint_indicatpor is TRUE, than the     #
# first endpoint is used otherwise the second  #
#--------------------------------------------#
if(endpoint_indicator){
datanew <- data[,c(2,3,5)]
}
```

```
else{
datanew <- data[,c(2,4,5)]
}


surv_info<-list(times=datanew[,1],
status=datanew[,2],
group=datanew[,3])

cox<-coxph(Surv(times,status)~group,surv_info)
teststat<- -(cox$coefficients-0)/(sqrt(cox$var))

if(sum(surv_info$status[(samplesize_I+1)
:(samplesize_I+samplesize_C)])==0
 |sum(surv_info$status[(1):(samplesize_I)])==0){
print("Warning: One or several bootstrap samples")
print("do not include any event within")
print("at least one of the treatment group")
print("Re-run to confirm correlation estimation.")
    }
      return(teststat)
    }

#-----------------------------------------------#
# For each bootstrap sample the test statistic  #
# for both endpoints are calculated through the #
# function "getTestStat"                        #
#-----------------------------------------------#

Teststat_ep <- mapply(getTestStat, sampleData)
Teststat_ce <- mapply(getTestStat, sampleData
,MoreArgs=list(endpoint_indicator=F))

#-----------------------------------------------#
# Calculate the correlation between the         #
# test statistics                               #
#-----------------------------------------------#

return(cor(Teststat_ce, Teststat_ep))
}


################################################
# Main function to calculate correlation dependent #
# parameter alpha_rho, the test statistics of both #
# endpoints and the test decision of the effect    #
# consistency approach (ECA)                       #
################################################


################################################
# Parameters to specify                        #
################################################
```

```
#
# Before you start the program it is necessary to
# import your data into R.
#
# Required structure of the data set "data":
#
# 1.column (id)       = Patient ID
# 2.column (times)    = Event time
# 3.column (status1)  = Binary event indicator for
#                       EP1 (Main Component)
# 4.column (status2)  = Binary event indicator for
#                       EP2 (Composite)
# 5.column (group)    = Group allocation
#---------------------------------------------------
# Other required parameters
#---------------------------------------------------
# alpha1    = Parameter of the ECA see Chap. 9.1.3
# alpha2    = Parameter of the ECA see Chap. 9.1.3
# alpha       = Global one-sided significance level
# alphaStern = Consistency parameter of the ECA
#             compare Chap. 9.1.3
# rho       = Correlation between the test statistics
#           under the intersection null hypothesis
###################################################

testdecision<-function(data,alpha1,alpha2,
                       alpha,alphaStern,rho){


  #-----------------------------------------------#
  # Calculate the test statistics of              #
  # EP1 (Main Component and EP2 (Composite)       #
  #-----------------------------------------------#

  surv_info<-list(times=data[,2],
                  status=data[,3],
                  group=data[,5])
  cox<-coxph(Surv(times,status)~group,surv_info)
  teststat1<- -(cox$coefficients-0)/(sqrt(cox$var))
  p1<-pnorm(teststat1)

  surv_info<-list(times=data[,2],
                  status=data[,4],
                  group=data[,5])
  cox<-coxph(Surv(times,status)~group,surv_info)
  teststat2<- -(cox$coefficients-0)/(sqrt(cox$var))
  p2<-pnorm(teststat2)


  #-----------------------------------------------#
  # Calculate alpha_rho by solving Eq.            #
  # (9.8) in Chap. 9 in dependence of rho         #
  #-----------------------------------------------#
```

```
r12<-matrix(c(1,rho,rho,1),ncol=2)
pow<-1
alphax<-0

pow1<-(pmvnorm(lower=c(-Inf,-Inf)
               ,upper=c(qnorm(1-alpha1)
                ,qnorm(1-alpha2))
               ,mean=c(0,0),corr=r12)
       -pmvnorm(lower=c(-Inf,-Inf)
                ,upper=c(qnorm(1-alpha)
                ,qnorm(1-alpha2))
                ,mean=c(0,0),corr=r12))

boarder <- alphaStern-alpha

while(pow>= boarder){
  alphax<-alphax + 0.001
  pow2<-(pmvnorm(lower=c(-Inf,-Inf)
          ,upper=c(qnorm(1-alpha)
          ,qnorm(1-alphax))
          ,mean=c(0,0),corr=r12)
         -pmvnorm(lower=c(-Inf,-Inf)
          ,upper=c(qnorm(1-alphaStern)
          ,qnorm(1-alphax))
          ,mean=c(0,0),corr=r12))
  pow<-pow1+pow2
}

while(pow>=boarder){
  alphax<-alphax-0.0001
  pow2<-(pmvnorm(lower=c(-Inf,-Inf)
          ,upper=c(qnorm(1-alpha)
          ,qnorm(1-alphax))
          ,mean=c(0,0), corr=r12)
         -pmvnorm(lower=c(-Inf,-Inf)
          ,upper=c(qnorm(1-alphaStern)
          ,qnorm(1-alphax))
          ,mean=c(0,0), corr=r12))
  pow<-pow1-pow2
}

alphax<-alphax+0.0001
alpha_rho<-min(alphax,alpha2)


#--------------------------------------------------------#
# Test decision of the ECA, compare Fig. 9.1 in Chap. 9  #
#--------------------------------------------------------#

if(teststat1>=qnorm(1-alpha1)
   && teststat2>=qnorm(1-alpha1)){
  print("Case 1: Reject both null hypotheses")
}
else if(teststat1>=qnorm(1-alpha1)
```

```
    && teststat2<qnorm(1-alpha1)){
    print("Case 2: Reject only the nullhypothesis")
    print("of EP1 (Main Component)")
  }
  else if(teststat1<qnorm(1-alpha1)
    && teststat1>=qnorm(1-alpha)
    && teststat2>=qnorm(1-alpha2)){
    print("Case 3: Reject both null hypotheses")
  }
  else if(teststat1<qnorm(1-alpha1)
    && teststat1>=qnorm(1-alpha)
    && teststat2<qnorm(1-alpha2)){
    print("Case 4: Keep both null hypotheses")
  }
  else if(teststat1<qnorm(1-alpha)
    && teststat1>=qnorm(1-alphaStern)
    && teststat2>=qnorm(1-alpha_rho)){
    print("Case 5: Reject the nullhypothesis")
    print(" of EP2 (Composite).")
    print("The effect of EP1 (Main Component)")
    print("is consistent.")
  }
  else if(teststat1<qnorm(1-alpha)
    && teststat1>=qnorm(1-alphaStern)
    && teststat2<qnorm(1-alpha_rho)){
    print("Case 6: Keep both null hypotheses")
  }
  else if(teststat1<qnorm(1-alphaStern) ){
    print("Case 7: Keep both null hypotheses")
  }

  print("Test statistics and p-values")
  print("of EP1 (Main Component),EP2 (Composite)")
  print(c(teststat1, 1-pnorm(teststat1)
         , teststat2,1-pnorm(teststat2)))
  print("Parameter alpha_rho")
  print(alpha_rho)
}

#************************************#
# Exemplary calls                    #
#************************************#
set.seed(150)

#------------------------------------#
# Generate data where each event in  #
# the first endpoint (main component) #


# also refers to an event in the     #
# second component (composite)       #
#------------------------------------#
```

```
#------------------------------------#
# Specify all-cause hazards          #
#------------------------------------#

haz_I<-0.5
haz_C<-0.8

#------------------------------------#
# Generate event indicators for two  #
# components                         #
#------------------------------------#

statusI<-rmultinom(40, 1,
        c(0.15/haz_I,0.1/haz_I
          ,1-(0.15+0.1)/haz_I))

statusC<-rmultinom(20, 1,
        c(0.32/haz_C,0.32/haz_C
          ,1-(0.32+0.32)/haz_C))


status1<-c(statusI[1,], statusC[1,])
status2h<-c(statusI[2,], statusC[2,])

#------------------------------------#
# Composite equals 1 whenever at     #
# least one of the components is 1   #
#------------------------------------#
status2<-rep(0,60)
status2[status2h==1 | status1==1] <-1


data<-data.frame(id=c(1:60),
     times=c(-log(runif(40,0,1))/(haz_I)
          ,-log(runif(20,0,1))/(haz_C)),
     status1=status1,
     status2=status2,
     group=c(rep(1,40),rep(0,20)))

print(data)
corr<-bootStrap_corr(data, 500)
print(corr)

testdecision(data,0.02,0.02,0.025,0.5,corr)
```

11.5 Adaptive Bonferroni and Bonferroni-Holm

11.5.1 Adaptive Bonferroni and Bonferroni-Holm for Binary Endpoints

```
library(mvtnorm)


###################################################
# Function to apply the group-sequential          #
# Bonferroni or Bonferroni-Holm procedure with    #
# with recalculation of                           #
# the number of required events                   #
# for an UIT based on two binary endpoints        #
#-------------------------------------------------#
# Local levels are adjusted according to Pocock   #
###################################################

###################################################
# Parameters to specify                           #
###################################################
#
#
# Before you start the program it is necessary to
# import your data into R.
#
# Required structure of the data set "data":
#
# 1.column (id)      = Patient ID
# 2.column (status1) = Binary event indicator for
#                      EP1 (Main Component)
# 3.column (status2) = Binary event indicator for
#                      EP2 (Composite)
# 4.column (group)   = Group allocation
#--------------------------------------------------
# Other required parameters
#--------------------------------------------------
# globalalpha  = Global one-sided signifcance level
# psi           = Information fraction at interim
# beta      = 1-beta is the anticipated power for
#             recalculation based on the
#             conditional power at interim
# procedure = "B" Bonferoni or "BH" Bonferoni Holm
###################################################


adaptiveUIT<-function(globalalpha, psi,
                      beta, procedure, data_interim){

  alpha1_half = alpha12_half = globalalpha/4
  alpha1_full = alpha12_full = globalalpha/2
```

```
#-------------------------------------------------#
# Calculate the corresponding group-sequential    #
# local significance levels for stage 1 and 2     #
# for the union-intersection test                 #
# based on the full and the half global level     #
# according to Eq. (10.1) and (10.4) in Sec. 10.2 #
#-------------------------------------------------#
sigma = matrix(c(1,sqrt(psi),sqrt(psi),1), ncol=2)

while(pmvnorm(lower=c(-Inf,-Inf),
              upper = c(qnorm(1-alpha1_half),
                        qnorm(1-alpha12_half)),
              mean  = c(0,0),
              sigma = sigma)[1]
      >= 1-globalalpha/2){

  alpha1_half   = alpha1_half + 0.0001
  alpha12_half  = alpha12_half + 0.0001

}

while(pmvnorm(lower=c(-Inf,-Inf),
              upper = c(qnorm(1-alpha1_full),
                        qnorm(1-alpha12_full)),
              mean  = c(0,0),
              sigma = sigma)[1]
      >= 1-globalalpha){
  alpha1_full    = alpha1_full + 0.0001
  alpha12_full   = alpha12_full + 0.0001
}

#------------------------------------------------#
# Calculate sample size of the groups            #
#------------------------------------------------#

samplesize_C =length(data[,1])-sum(data[,4])
samplesize_I =sum(data[,4])
n1<-samplesize_C+samplesize_I

#------------------------------------------------#
# Calculate the test statistics of               #
# EP1 (Main Component) and EP2 (Composite)       #
#------------------------------------------------#

data_ep1 = data[,c(2,4)]
data_ep2 = data[,c(3,4)]

data_ep1_C = data_ep1[data_ep2[,2] == 0,]
data_ep1_I = data_ep1[data_ep2[,2] == 1,]

data_ep2_C = data_ep2[data_ep2[,2] == 0,]
data_ep2_I = data_ep2[data_ep2[,2] == 1,]

p_ep1_C = mean(data_ep1_C[,1])
```

```
p_ep1_I = mean(data_ep1_I[,1])
p_ep1 = (p_ep1_C*samplesize_C+
          p_ep1_I*samplesize_I)/
  (samplesize_C+samplesize_C)
var_ep1_0 = p_ep1*(1-p_ep1)*
  (samplesize_C+samplesize_C)/
  (samplesize_C*samplesize_C)
teststat1 = (p_ep1_C-p_ep1_I)/sqrt(var_ep1_0)

p_ep2_C = mean(data_ep2_C[,1])
p_ep2_I = mean(data_ep2_I[,1])
p_ep2 = (p_ep2_C*samplesize_C+
          p_ep2_I*samplesize_I)/
  (samplesize_C+samplesize_C)
var_ep2_0 = p_ep2*(1-p_ep2)*
  (samplesize_C+samplesize_C)/
  (samplesize_C*samplesize_C)
teststat2 = (p_ep2_C-p_ep2_I)/sqrt(var_ep2_0)

#-----------------------------------------------#
# Choose the most promising endpoint to         #
# recalculate the sample size                   #
#-----------------------------------------------#

if(teststat1>teststat2) {
  promise   = 1
}else{
  promise   = 2
}

#------------------------------------------------#
# Apply Bonferroni or Bonferroni-Holm procedure  #
# for interim test decision                      #
#------------------------------------------------#

if(procedure=="B"){
  print("Bonferroni")
  if(teststat1>=qnorm(1-alpha1_half)|
     teststat2>=qnorm(1-alpha1_half)){
    if(teststat1>=qnorm(1-alpha1_half)&
       teststat2<qnorm(1-alpha1_half)){
      ind<-"EP1"}
    else if(teststat1<qnorm(1-alpha1_half)&
            teststat2>=qnorm(1-alpha1_half)){
      ind<-"EP2"}
    else if(teststat1>=qnorm(1-alpha1_half)&
            teststat2>=qnorm(1-alpha1_half)){
      ind<-"EP1 and EP2"}
    print(paste("Study stops at interim",
                "with rejection of",
                ind,
                "at level alpha_(1,alpha/2)",
                alpha1_half,
                ". P-values for EP1 and EP2 are",
```

```
                    round(1-pnorm(teststat1),4),"and",
                    round(1-pnorm(teststat2),4),
                    ". The most",
                    "promising endpoint is EP",
                    promise))
      }else{
        n2 = n1*((qnorm(beta)-
                qnorm(1-alpha12_half)*1/sqrt(1-psi))/
                   max(teststat1, teststat2)+
                   sqrt(psi)/sqrt(1-psi))^2
        print(paste("Continue study:",
         "The local level at interim alpha1_half is",
                    alpha1_half,
                    ". P-values for EP1 and EP2 are",
                    round(1-pnorm(teststat1),4),"and",
                    round(1-pnorm(teststat2),4),
                    ". The most",
                    "promising endpoint is EP",
                    promise,
                    ". The recalculated ",
                    "number of patients for the",
                    "second stage for this endpoint is",
                    ceiling(n2),
                    ". A number of",
                    n1,
                    "patients are already",
                    "observed at interim."))
      }
    }else{
      print("Bonferron-Holm")
      if(teststat1>=qnorm(1-alpha1_half)|
         (((qnorm(1-alpha1_half)>=teststat1)&
           (teststat1>qnorm(1-alpha1_full)))
          &teststat2>=qnorm(1-alpha1_half))|
         (teststat2>=qnorm(1-alpha1_half)|
          (((qnorm(1-alpha1_half)>=teststat2)&
            (teststat2>qnorm(1-alpha1_full)))
           &teststat1>=qnorm(1-alpha1_half))
         )) {
        if(teststat1>=qnorm(1-alpha1_half)&
           teststat2<qnorm(1-alpha1_full)){ind<-"EP1"}
        else if(teststat1<qnorm(1-alpha1_full)&
                teststat2>=qnorm(1-alpha1_half)){
          ind<-"EP2"}
        else if((teststat1>=qnorm(1-alpha1_half)&
                 teststat2>=qnorm(1-alpha1_full))|
                (teststat1>=qnorm(1-alpha1_full)&
                 teststat2>=qnorm(1-alpha1_half))){
          ind<-"EP1 and EP2"}
        print(paste("Study stops at interim",
                    "with rejection of",
                    ind,
      "at levels alpha_(1,alpha/2) and alpha_(1,alpha)",
                    alpha1_half, alpha1_full,
```

```
                    ". P-values for EP1 and EP2 are",
                    round(1-pnorm(teststat1),4),"and",
                    round(1-pnorm(teststat2),4),
                    ". The most",
                    "promising endpoint is EP",
                    promise))
      }else{
          n2    = n1*((qnorm(beta)-
                 qnorm(1-alpha12_half)*1/sqrt(1-psi))/
                     max(teststat1, teststat2)+
                     sqrt(psi)/sqrt(1-psi))^2
        print(paste("Continue study:",
                    "The local levels at interim",
                    "alpha1_half and alpha1_full are",
                    alpha1_half, alpha1_full,
                    ". P-values for EP1 and EP2 are",
                    round(1-pnorm(teststat1),4),"and",
                    round(1-pnorm(teststat2),4),
                    ". The most",
                    "promising endpoint is EP",
                    promise,
                    ". The recalculated ",
                    "number of patients for the",
                  "second stage for this endpoint is",
                    ceiling(n2),
                    ". A number of",
                    n1,
                    "patients are already",
                    "observed at interim."))
      }
    }

}
#************************************#
# Exemplary Calls                    #
#************************************#
set.seed(120)

#------------------------------------#
# Generate data where each event in  #
# the first endpoint (main component) #
# also refers to an event in the     #
# second component (composite)       #
#------------------------------------#

#------------------------------------#
# Generate event indicators for two  #
# components                         #
#------------------------------------#

statusI<-rmultinom(80, 1, c(0.1,0.2,0.7))
statusC<-rmultinom(80, 1, c(0.25,0.25,0.5))
```

```
status1<-c(statusI[1,], statusC[1,])
status2h<-c(statusI[2,], statusC[2,])

#-------------------------------------#
# Composite equals 1 whenever at      #
# least one of the components is 1    #
#-------------------------------------#

status2<-rep(0,160)
status2[status2h==1 | status1==1] <-1

data<-data.frame(id=c(1:160),
                 status1=status1,
                 status2=status2,
                 group=c(rep(1,80),rep(0,80)))
print(data)


adaptiveUIT(0.025, 0.5, 0.2, "B", data)
adaptiveUIT(0.025, 0.5, 0.2, "BH", data)
#----------------------------------------
adaptiveUIT(0.05, 0.5, 0.2, "B", data)
adaptiveUIT(0.05, 0.5, 0.2, "BH", data)
```

11.5.2 Adaptive Bonferroni and Bonferroni-Holm for Time-to-Event Endpoints

```
library(mvtnorm)
library(survival)

##################################################
# Function to apply the group-sequential         #
# Bonferroni or Bonferroni-Holm procedure with   #
# with recalculation of                          #
# the number of required events                  #
# for an UIT based on two time-to-event endpoints #
#------------------------------------------------#
# Local levels are adjusted according to Pocock  #
##################################################

##################################################
# Parameters to specify                          #
##################################################
#
#
# Before you start the program it is necessary to
# import your interim data into R. The data set
# "data_interim" thus corresponds to the observed
# data at the interim stage.
#
```

```
# Required structure of the data set "data_interim":
#
# 1.column (id)      = Patient ID
# 2.column (times)   = Event time
# 3.column (status1) = Binary event indicator for
#                      EP1 (Main Component)
# 4.column (status2) = Binary event indicator for
#                      EP2 (Composite)
# 5.column (group)   = Group allocation
#---------------------------------------------------
# Other required parameters
#---------------------------------------------------
# globalalpha  = Global one-sided signifcance level
# psi          = Information fraction at interim
# beta      = 1-beta is the anticipated power for
#             recalculation based on the
#             conditional power at interim
# procedure = "B" Bonferoni or "BH" Bonferoni Holm
###################################################


adaptiveUIT<-function(globalalpha, psi,
                      beta, procedure, data_interim){

  alpha1_half = alpha12_half = globalalpha/4
  alpha1_full = alpha12_full = globalalpha/2

  #-----------------------------------------------#
  # Calculate the corresponding group-sequential  #
  # local significance levels for stage 1 and 2   #
  # for the union-intersection test               #
  # based on the full and the half global level   #
  # according to Eq. (10.1) and (10.4) in Sec. 10.2 #
  #-----------------------------------------------#
  sigma = matrix(c(1,sqrt(psi),sqrt(psi),1), ncol=2)

  while(pmvnorm(lower=c(-Inf,-Inf),
              upper = c(qnorm(1-alpha1_half),
                        qnorm(1-alpha12_half)),
              mean  = c(0,0),
              sigma = sigma)[1]
        >= 1-globalalpha/2){

    alpha1_half   = alpha1_half + 0.0001
    alpha12_half  = alpha12_half + 0.0001

  }


  while(pmvnorm(lower=c(-Inf,-Inf),
              upper = c(qnorm(1-alpha1_full),
                        qnorm(1-alpha12_full)),
              mean  = c(0,0),
```

```
                sigma = sigma)[1]
        >= 1-globalalpha){
    alpha1_full     = alpha1_full + 0.0001
    alpha12_full    = alpha12_full + 0.0001
}

#-------------------------------------------------#
# Calculate Wald-type test statistics of both     #
# endpoint at interim                             #
#-------------------------------------------------#

cox1      = coxph(Surv(times,status1)~group,data)
teststat1 = -(cox1$coefficients-0)/(sqrt(cox1$var))
cox2      = coxph(Surv(times,status2)~group,data)
teststat2 = -(cox2$coefficients-0)/(sqrt(cox2$var))

#-------------------------------------------------#
# Choose the most promising endpoint to           #
# recalculate the number of events needed         #
#-------------------------------------------------#

if(teststat1>teststat2) {
  d1        = cox1$nevent
  promise   = 1
}else{
  d1        = cox2$nevent
  promise   = 2
}

#-------------------------------------------------#
# Apply Bonferroni or Bonferroni-Holm procedure   #
# for interim test decision                       #
#-------------------------------------------------#

if(procedure=="B"){
  print("Bonferroni")
  if(teststat1>=qnorm(1-alpha1_half)|
     teststat2>=qnorm(1-alpha1_half)){
    if(teststat1>=qnorm(1-alpha1_half)&
       teststat2<qnorm(1-alpha1_half)){ind<-"EP1"}
    else if(teststat1<qnorm(1-alpha1_half)&
            teststat2>=qnorm(1-alpha1_half)){
      ind<-"EP2"}
    else if(teststat1>=qnorm(1-alpha1_half)&
            teststat2>=qnorm(1-alpha1_half)){
      ind<-"EP1 and EP2"}
    print(paste("Study stops at interim",
                "with rejection of",
                ind,
                "at level alpha_(1,alpha/2)",
                alpha1_half,
                ". P-values for EP1 and EP2 are",
                round(1-pnorm(teststat1),4),"and",
                round(1-pnorm(teststat2),4),
```

```
                ". The most",
                "promising endpoint is EP",
                promise))
    }else{
      d2 = d1*((qnorm(beta)-
               qnorm(1-alpha12_half)*1/sqrt(1-psi))/
               max(teststat1, teststat2)+
               sqrt(psi)/sqrt(1-psi))^2
      print(paste("Continue study:",
         "The local level at interim alpha1_half is",
                alpha1_half,
                ". P-values for EP1 and EP2 are",
                round(1-pnorm(teststat1),4),"and",
                round(1-pnorm(teststat2),4),
                ". The most",
                "promising endpoint is EP",
                promise,
                ". The recalculated ",
                "number of events for the",
                "second stage for this endpoint is",
                ceiling(d2),
                ". A number of",
                d1,
                "events are already",
                "observed at interim."))
    }
  }else{
    print("Bonferron-Holm")
    if(teststat1>=qnorm(1-alpha1_half)|
       (((qnorm(1-alpha1_half)>=teststat1)&
         (teststat1>qnorm(1-alpha1_full)))
        &teststat2>=qnorm(1-alpha1_half))|
       (teststat2>=qnorm(1-alpha1_half)|
        (((qnorm(1-alpha1_half)>=teststat2)&
          (teststat2>qnorm(1-alpha1_full)))
         &teststat1>=qnorm(1-alpha1_half))
       )) {
      if(teststat1>=qnorm(1-alpha1_half)&
         teststat2<qnorm(1-alpha1_full)){
        ind<-"EP1"}
      else if(teststat1<qnorm(1-alpha1_full)&
              teststat2>=qnorm(1-alpha1_half)){
        ind<-"EP2"}
      else if((teststat1>=qnorm(1-alpha1_half)&
               teststat2>=qnorm(1-alpha1_full))|
              (teststat1>=qnorm(1-alpha1_full)&
               teststat2>=qnorm(1-alpha1_half))){
        ind<-"EP1 and EP2"}
      print(paste("Study stops at interim",
                "with rejection of",
                ind,
  "at levels alpha_(1,alpha/2) and alpha_(1,alpha)",
                alpha1_half, alpha1_full,
                ". P-values for EP1 and EP2 are",
```

```
                    round(1-pnorm(teststat1),4),"and",
                    round(1-pnorm(teststat2),4),
                    ". The most",
                    "promising endpoint is EP",
                    promise))
    }else{
      d2    = d1*((qnorm(beta)-
                  qnorm(1-alpha12_half)*1/sqrt(1-psi))/
                  max(teststat1, teststat2)+
                  sqrt(psi)/sqrt(1-psi))^2
      print(paste("Continue study:",
                    "The local levels at interim",
                    "alpha1_half and alpha1_full are",
                    alpha1_half, alpha1_full,
                    ". P-values for EP1 and EP2 are",
                    round(1-pnorm(teststat1),4),"and",
                    round(1-pnorm(teststat2),4),
                    ". The most",
                    "promising endpoint is EP",
                    promise,
                    ". The recalculated ",
                    "number of events for the",
                    "second stage for this endpoint is",
                    ceiling(d2),
                    ". A number of",
                    d1,
                    "events are already",
                    "observed at interim."))
    }
  }


}

#**********************************#
# Exemplary Calls                  #
#**********************************#
set.seed(120)

#----------------------------------#
# Generate data where each event in  #
# the first endpoint (main component) #
# also refers to an event in the   #
# second component (composite)     #
#----------------------------------#

#----------------------------------#
# Specify all-cause hazards        #
#----------------------------------#

haz_I<-0.7
haz_C<-0.8
```

```
#-------------------------------------#
# Generate event indicators for two   #
# components                          #
#-------------------------------------#

statusI<-rmultinom(100, 1,
                   c(0.5/haz_I,0.1/haz_I
                     ,1-(0.5+0.1)/haz_I))
statusC<-rmultinom(100, 1,
                   c(0.5/haz_C,0.3/haz_C
                     ,1-(0.5+0.3)/haz_C))


status1<-c(statusI[1,], statusC[1,])
status2h<-c(statusI[2,], statusC[2,])

#-------------------------------------#
# Composite equals 1 whenever at      #
# least one of the components is 1    #
#-------------------------------------#
status2<-rep(0,200)
status2[status2h==1 | status1==1] <-1


data<-data.frame(id=c(1:200),
                 times=c(-log(runif(100,0,1))/(haz_I)
                      ,-log(runif(100,0,1))/(haz_C)),
                 status1=status1,
                 status2=status2,
                 group=c(rep(1,100),rep(0,100)))


adaptiveUIT(0.025, 0.5, 0.2, "B", data)
adaptiveUIT(0.025, 0.5, 0.2, "BH", data)
#----------------------------------------
adaptiveUIT(0.05, 0.5, 0.2, "B", data)
adaptiveUIT(0.05, 0.5, 0.2, "BH", data)
```

Part IV
Confirmatory Test Problem for a Weighted Composite Endpoint

A completely different approach to ease the interpretation of a composite endpoint without formulating a multiple test problem is to directly define a weighted composite effect measure, where the weights reflect the clinical relevance of the different components. By this, a treatment effect in the composite can no longer be caused by large effects in components of low clinical relevance. Of course, even with a weighting approach, the treatment effect of the composite allows no conclusion on the treatment effects of the single components, however, this issue is less problematical if the composite effect truly reflects the effect of the intervention under investigation. Efficacy is shown whenever the intervention is significantly superior to the control with respect to the weighted effect.

A desirable property of a weighted effect measure is that it defines a natural extension of the commonly used unweighted effect measure to simplify the interpretation. Several authors have proposed weighting strategies for composite endpoints which do not all meet this criterion. In this part of the book, we will present several exemplary weighting approaches and discuss their properties and related problems. Moreover, we will deduce recommendations for application of the different methods in practice.

Chapter 12
Weighted Composite Binary Endpoint

A variety of weighted effect measures and analysis approaches for composite endpoints were proposed in the literature of which only a few are applicable to composite binary endpoints (Pocock et al. 2012; Buyse 2010). Many of these approaches should be applied with care as explored in more detail in Chap. 14. Therefore, in this chapter, an alternative weighted effect measure for a composite binary endpoint will be introduced. As motivated above, an intuitive way to define a weighted effect measure is to extend the standard unweighted effect measure in a reasonable way. For a binary endpoint, the common effect measure is the absolute risk difference. In the remainder of this chapter, a weighted version of this risk difference will be introduced.

12.1 Weighted Risk Difference

In order to define a weighted effect measure for composite binary endpoints, the notations introduced in Chap. 5 of Part II will be used in the following. As mentioned above, a standard unweighted effect measure for composite binary endpoints is given by the absolute risk difference

$$\Delta_{CE} := p^I_{CE} - p^C_{CE} = \sum_{l=1}^{k} \left(p^I_{EP_l,ex} - p^C_{EP_l,ex}\right) + \sum_{j=1}^{m} \left(p^I_{\cap j} - p^C_{\cap j}\right), \qquad (12.1)$$

where

$$m := \sum_{l=2}^{k} \binom{k}{l}.$$

G. Rauch et al., *Planning and Analyzing Clinical Trials with Composite Endpoints*, Springer Series in Pharmaceutical Statistics,
https://doi.org/10.1007/978-3-319-73770-6_12

Weighted composite event rates can be easily defined by assigning weighting factors to the different components and overlap event rates. Therefore, let

$$w_l \geq 0, l = 1, \ldots, k, \tag{12.2}$$

and

$$w_{\cap_j}, j = 1, \ldots, m, \tag{12.3}$$

correspond to some predefined weights fulfilling

$$\sum_{l=1}^{k} w_l + \sum_{j=1}^{m} w_{\cap_j} = 1. \tag{12.4}$$

Note that this means that separate weights have to be defined for all unique event types and for all overlap combinations. Then, the weighted composite event rates are given as follows

$$p_{CE}^{I,w} := \sum_{l=1}^{k} w_l \cdot p_{EP_l,ex}^{I} + \sum_{j=1}^{m} w_{\cap_j} \cdot p_{\cap j}^{I},$$

$$p_{CE}^{C,w} := \sum_{l=1}^{k} w_l \cdot p_{EP_l,ex}^{C} + \sum_{j=1}^{m} w_{\cap_j} \cdot p_{\cap j}^{C},$$

where w corresponds to the vector of all weights introduced in Eqs. (12.2) and (12.3). Using these notations, a weighted effect measure for composite binary endpoints can be defined which will be referred as the "weighted risk difference" in the following:

$$\Delta_{CE}^{w} := \sum_{l=1}^{k} w_l \left(p_{EP_l,ex}^{I} - p_{EP_l,ex}^{C}\right) + \sum_{j=1}^{m} w_{\cap_j} \left(p_{\cap_j}^{I} - p_{\cap_j}^{C}\right). \tag{12.5}$$

For the special case of mutually exclusive components which means that $p_{\cap_j}^{I} = p_{\cap_j}^{C} = 0,\ j = 1, \ldots, m$, the weighted composite effect measure can also be written as

$$\Delta_{CE}^{w} := \sum_{l=1}^{k} w_l \left(p_{EP_l}^{I} - p_{EP_l}^{C}\right), \tag{12.6}$$

where the weights $w_l \geq 0,\ l = 1, \ldots, k$, satisfy

$$\sum_{l=1}^{k} w_l = 1. \tag{12.7}$$

Note that for the special case of equal weights across the components and overlap rates, the weighted risk difference corresponds to the standard risk difference. Thus, the weighted risk difference defines a natural extension of the common unweighted approach.

12.1.1 Considerations on the Choice of Weights

An important aspect of an adequate and meaningful analysis using a weighted composite binary effect measure is that the weights w_l, $l = 1, \ldots, k$, and $w_{\cap j}, j = 1, \ldots, m$, must be prespecified in the planning stage. In principle, the weights can be chosen arbitrarily. The magnitude of the weights should thereby reflect the relative importance of the components to each other. Most harmful events such as different causes of death should usually correspond to large weights. A rule of thumb could be that the sum of the weights for all fatal events should be at least 0.5. The set of components of minor relevance should only comprise a weight sum of at most 0.2, whereas events of intermediate relevance should be represented by weights in between these values. When specifying the weights for the overlap rates, the magnitude of the weights can be guided by the weight of the most relevant component contained within the overlap. However, it should also be noted that the number of patients experiencing a specific overlap combination of events will usually be limited and therefore the corresponding weights should not be too high. All these considerations are of very general nature which might help to choose adequate weights but cannot be considered as general recommendation because the specific clinical trial situation always has to be taken into account.

12.1.2 Test Problem

The test hypotheses to be assessed in the confirmatory analysis in terms of the weighted composite effect are given as

$$H_0^{CE} : \Delta_{CE}^w \leq 0 \text{ versus } H_1^{CE} : \Delta_{CE}^w > 0. \tag{12.8}$$

12.1.3 Test Statistic

A statistical test to assess the above null hypothesis for the weighted composite effect can be defined similarly as the normal approximation test for rates for the standard unweighted approach.

$$T_{CE}^w := \frac{\widehat{\Delta}_{CE}^w}{\widehat{\sigma}_{\Delta_{CE}^w}}. \tag{12.9}$$

This test will in the following be referred as the "weighted normal approximation test for rates." Thereby, $\widehat{\Delta}^w_{CE}$ is given by plugging in the maximum likelihood estimators for the disjoint event probabilities $p^I_{EP_l,ex}, p^C_{EP_l,ex}$, $l = 1,\ldots,k$, and $p^I_{\cap_j}, p^C_{\cap_j}$, $j = 1,\ldots,m$, given as the corresponding relative frequencies. The restricted maximum likelihood estimator of the treatment effect variance under the null hypothesis restriction $\Delta^w_{CE} = 0$ is given by

$$\widehat{\sigma}^2_{\Delta^w_{CE}} = \widehat{p}^w_{CE} \cdot \left(1 - \widehat{p}^w_{CE}\right) \cdot \frac{n_I + n_C}{n_I \cdot n_C}, \tag{12.10}$$

where

$$\widehat{p}^w_{CE} = \frac{\widehat{p}^{I,w}_{CE} \cdot n_I + \widehat{p}^{C,w}_{CE} \cdot n_C}{n_I + n_C}.$$

Using these notations, the test statistic (12.9) is approximately distributed as

$$T^w_{CE} \sim N\left(\frac{\Delta^w_{CE}}{\sigma_{\Delta^w_{CE}}}, 1\right) \tag{12.11}$$

Under the corresponding null hypothesis (12.8), the test statistic T^w_{CE} thus asymptotically follows a standard normal distribution. Therefore, the null hypothesis is rejected whenever $T^w_{CE} \geq z_{1-\alpha}$, where α denotes the one-sided significance level.

12.1.4 Sample Size Calculation

The derivation of the sample size formula is equivalent as for the normal approximation test for rates described in Sect. 5.1.3.1. Thus, the total sample size for both groups is given by the smallest value of the total sample size fulfilling

$$n_I + n_C \geq \left(1 + \frac{1}{\gamma}\right) \cdot \left(\frac{z_{1-\alpha}\sqrt{(\gamma+1)\, p^{C,w}_{CE}\left(1 - p^{C,w}_{CE}\right)} + z_{1-\beta}\sqrt{p^{I,w}_{CE}(1-p^{I,w}_{CE}) + \gamma p^{C,w}_{CE}(1-p^{C,w}_{CE})}}{\Delta^w_{CE}}\right)^2$$

with allocation ratio of sample size $\gamma = n_I/n_C$. Note, however, that the calculation of sample size requires valid estimates of $p^{I,w}_{CE}, p^{C,w}_{CE}$ as well as of the weighted treatment effect Δ^w_{CE}. Valid assumptions on these parameters are difficult as this requires the correct specification of all underlying event and overlap rates. Therefore, we recommend to investigate the robustness of the sample size, for example, by calculating the worst-case sample size based on a range of plausible scenarios. The weighted normal approximation test for rates can also be used within a group sequential or an adaptive design which are described in Chap. 6 in Part II of this book. All mathematical considerations are equivalent to the standard unweighted version.

References

Buyse, M. (2010). Generalized pairwise comparisons of prioritized outcomes in the two-sample problem. *Statistics in Medicine, 29*, 3245–3257.

Pocock, S. J., Ariti, C. A., Collier, T. J., & Wang, D. (2012). The win ratio: A new approach to the analysis of composite endpoints in clinical trials based on clinical priorities. *European Heart Journal, 33*, 176–182.

Chapter 13
Weighted Composite Time-to-Event Endpoint

A variety of weighted effect measures and analysis approaches for composite endpoints were proposed in the literature of which most refer to time-to-event composite endpoints (Lachin and Bebu 2015; Bakal et al. 2015, 2013; Pocock et al. 2012; Buyse 2010; Hallstrom et al. 1992). Similar as for composite binary endpoints, application of most methods should be done with care as will be explored in more detail in Chap. 14, compare also Bebu and Lachin (2016) and Rauch et al. (2014). In this chapter, we introduce a weighted effect measure and a related test as proposed by Rauch et al. (2018). The common effect measure for composite time-to-event endpoints is the all-cause hazard ratio which gives equal weight to all types of events irrespective of their clinical relevance. Thereby, the all-cause hazard within each group is given by the sum of the cause-specific hazards corresponding to the individual components. A natural extension of the standard all-cause hazard ratio can be defined by a "weighted all-cause hazard ratio" where the individual hazards for each component are multiplied with predefined relevance weighting factors. For the special case of equal weights across the components, the weighted all-cause hazard ratio then corresponds to the standard all-cause hazard ratio. In this chapter, the weighted all-cause hazard ratio will be introduced as a new weighted effect measure. Moreover, a related statistical test will be proposed.

13.1 Weighted All-Cause Hazard Ratio

Using the notations introduced in Chap. 5 in Part II of this book, a weighted effect measure for time-to-event composite endpoints can be defined which is referred as the "weighted all-cause hazard ratio"; compare Rauch et al. (2018)

$$\theta_{CE}^{w}(t) = \frac{\sum_{l=1}^{k} w_l \cdot \lambda_{EP_l}^{C}(t)}{\sum_{l=1}^{k} w_l \cdot \lambda_{EP_l}^{I}(t)}, \tag{13.1}$$

where w corresponds to the vector of all weights $w_l \geq 0,\ l = 1, \ldots, k$, which are given as nonnegative weighting factors reflecting the clinical relevance of EP_j.

G. Rauch et al., *Planning and Analyzing Clinical Trials with Composite Endpoints*, Springer Series in Pharmaceutical Statistics,
https://doi.org/10.1007/978-3-319-73770-6_13

For equal weights $w_1 = w_2 = \ldots = w_k$, the weighted all-cause hazard ratio corresponds to the standard all-cause hazard ratio. This property is appealing, as the weighted effect measure thus defines a natural extension of the commonly applied all-cause hazard ratio. The weighted all-cause hazard ratio is a time-dependent effect measure as long as the cause-specific hazards are nonconstant. It can thus either be reported for a predefined reasonable time point t or, as an alternative, might be averaged over the complete observational period

$$\Theta^w_{CE}(f) := \frac{1}{f}\int_0^f \theta^w_{CE}(t), \tag{13.2}$$

where $[0;f]$ defines the observational interval.

Estimators for $\theta^w_{CE}(t)$ and $\Theta^w_{CE}(f)$ can be obtained by plugging in estimators for the underlying cause-specific hazards. The latter can be obtained from any parametric survival model. Let $\widehat{\lambda}^I_{EP_j}$ and $\widehat{\lambda}^C_{EP_j}$ denote the corresponding estimates of the hazard functions resulting from the selected survival models, then an estimator for $\theta^w_{CE}(t)$ is given by

$$\widehat{\theta}^w_{CE}(t) = \frac{\sum_{l=1}^k w_l \cdot \widehat{\lambda}^C_{EP_l}(t)}{\sum_{l=1}^k w_l \cdot \widehat{\lambda}^I_{EP_l}(t)}. \tag{13.3}$$

The estimator (13.3) is consistent as the nominator and the denominator of (13.3) are both continuous functions of consistently estimated parameters which are thus also consistent by the continuous mapping theorem. As the denominator converges to a constant, applying Slutky's theorem yields the consistence of (13.3); compare Lehmann (1999). A consistent estimator $\widehat{\Theta}^w_{CE}(f)$ can be defined equivalently.

13.1.1 Considerations on the Choice of Weights

An important aspect of an adequate and meaningful analysis using a weighted composite effect measure is that the weights must be prespecified in the planning stage. In principle, the weighting factors $w_l > 0,\ l = 1, \ldots, k$, can be chosen freely without any restriction. The magnitude of these factors should thereby reflect the relative importance of the components to each other. Note that the influence of an individual component on the weighted effect measure becomes higher if the corresponding weight is large but also if the underlying cause-specific hazard is large. Therefore, components referring to higher hazards are naturally up-weighted. A very small hazard caused by a low number of events will thus only influence the overall weighted effect if the relevance weighting factor is extremely large. In general, the largest weights should be assigned to the most harmful events (e.g., fatal events). However, the specific clinical trial situation has always to be taken into account.

13.1.2 Test Problem

The test problem to be assessed in the confirmatory analysis in terms of the weighted all-cause hazard ratio is given by

$$H_0^{CE} : \theta_{CE}^w(t) \leq 1 \text{ versus } H_1^{CE} : \theta_{CE}^w(t) > 1. \quad (13.4)$$

Equivalently, the test hypotheses can be formulated in terms of the integrated weighted all-cause hazard ratio

$$H_0^{CE} : \Theta_{CE}^w(f) \leq 1 \text{ versus } H_1^{CE} : \Theta_{CE}^w(f) > 1. \quad (13.5)$$

13.1.3 Test Statistic

A potential drawback of our approach is that the variance of the estimators $\widehat{\theta}_{CE}^w(t)$ and $\widehat{\Theta}_{CE}^w(f)$ cannot directly be deduced, and therefore the asymptotic distributions are unknown. In order to overcome this problem, the above hypotheses can be assessed by means of a permutation test. The `R` source code to calculate the (averaged) weighted all-cause hazard ratio and the corresponding permutation test is provided in Chap. 15, so application in practice is straightforward.

13.1.4 Sample Size Calculation

The derivation of the sample size is difficult as the power of the weighted all-cause hazard ratio test depends on the underlying cause-specific hazard ratios, on the expected number of events for each event type, and on the expected proportion of events within the observational period. Furthermore, the analysis is performed by a permutation test, and therefore no closed sample size formula is available. The sample size can thus only be approximated by simulations. For this purpose, the `R` code given in Chap. 15 to calculate the (averaged) weighted all-cause hazard ratio and the corresponding permutation test might be used.

13.1.5 Possible Extensions

The (averaged) weighted all-cause hazard ratio can also be estimated based on covariate-adjusted estimates of the cause-specific hazards. It is also possible to include hazard estimates for the transition between a first and a second event or between any subsequent events. This has the advantage not to waive the information resulting from subsequent events occurring after the first one. Unless for large cohort studies, the sample size of an interventional randomized controlled trial is usually limited, and thus the number of subsequent events is also limited, especially when differentiated between the various event types. As a consequence, the corresponding transition hazards will be small, and therefore their impact on the magnitude of the (averaged) weighted all-cause hazard ratio is usually limited.

13.2 Related Approaches

Lachin and Bebu (2015) proposed a related weighted effect measure given as a weighted average of the log-transformed component hazard ratios. In their original approach, the component hazard ratios are not given by the cause-specific hazard ratios but by the individual hazard ratio, which ignores the competing risk scenario. However, their approach can easily be adapted to the cause-specific hazard ratios to account for competing risks. An important advantage of the approach proposed by Lachin and Bebu (2015) is that the distribution of the effect measure and of the related test statistic is known, and thus a parametric statistical test can be deduced. Moreover, as only estimators of the component hazard ratios are required, the baseline hazards must not be identified, and thus no parametric survival model is required.

A potential drawback of the latter method is that the weighted average of the log-transformed cause-specific hazards is not directly related to the common all-cause hazard ratio. Moreover, by weighting the log-transformed cause-specific hazard ratios, the number of events for each component does not influence the magnitude of the weighted effect measure but only its variance. By this, the weighted effect can be driven by a large effect in an individual component even if the component effect is highly variable due to a small event number. Finally, as the variance of the weighted average of the log-transformed cause-specific hazards is basically a weighted sum of the individual component variances, the variance increases with increasing number of components. In contrast, a composite endpoint combines several components with the aim to increase the number of events and thereby reducing the variance.

Nonetheless, the weighted effect measure by Lachin and Bebu (2015) can be seen as an attractive alternative to the weighted all-cause hazard ratio presented above. The advantages and disadvantages of both approaches should thus be carefully considered when defining an analysis strategy for a specific clinical trial application.

References

Bakal, J. A., Westerhout, C. M., & Armstrong, P. W. (2015). Impact of weighted composite compared to traditional composite endpoints for the design of randomized controlled trials. *Statistical Methods in Medical Research, 24*, 980–988.

Bakal, J. A., Westerhout, C. M., Cantor, W. J., Fernández-Avilés, F., Welsh, R. C., Fitchett, D., et al. (2013). Evaluation of early percutaneous coronary intervention vs. standard therapy after fibrinolysis for ST-segment elevation myocardial infarction: Contribution of weighting the composite endpoint. *European Heart Journal, 34*, 903–908.

Bebu, I., & Lachin, J. M. (2016). Large sample inference for a win ratio analysis of a composite outcome based on prioritized outcomes. *Biostatistics, 17*, 178–187.

Buyse, M. (2010). Generalized pairwise comparisons of prioritized outcomes in the two-sample problem. *Statistics in Medicine, 29*, 3245–3257.

Hallstrom, A. P., Litwin, P. E., & Weaver, W. D. (1992). A method of assigning scores to the components of a composite outcome: An example from the MITI trial. *Controlled Clinical Trials, 13*, 148–155.

Lachin, J. M., & Bebu, I. (2015). Application of the Wei-Lachin multivariate one-directional test to multiple event-time outcomes. *Clinical Trials, 12*, 627–633.

Lehmann, E. L. (1999). *Elements of large-sample theory*. New York: Springer.

Pocock, S. J., Ariti, C. A., Collier, T. J., & Wang, D. (2012). The win ratio: A new approach to the analysis of composite endpoints in clinical trials based on clinical priorities. *European Heart Journal, 33*, 176–182.

Rauch, G., Jahn-Eimermacher, A., Brannath, W., & Kieser, M. (2014). Opportunities and challenges of combined effect measures based on prioritized outcomes. *Statistics in Medicine, 33*, 1104–1120.

Rauch, G., Kunzmann, K., Kieser, M., Wegscheider, K., & Eulenburg, C. (2018). A weighted combined effect measure for the analysis of a composite time-to-first-event endpoint with components of different clinical relevance. *Statistics in Medicine, 37*, 749–767.

Chapter 14
Other Weighted Effect Measures

A number of other effect measures and analysis approaches for composite endpoints which take account of the different clinical relevance of the components were proposed in the literature; compare, for example, Lachin and Bebu (2015), Bakal et al. (2015, 2013), Pocock et al. (2012), Buyse (2010), and Hallstrom et al. (1992). As it goes beyond the scope of this book to present a complete and systematic literature review on weighted approaches, only a few exemplary methods will be presented and discussed within this chapter. Many of the weighting approaches presented in the literature should be used and interpreted with care as they are subject to general mathematical concerns. For example, unlike the weighting approaches presented in the previous chapters, many of the approaches proposed in the literature combine first and subsequent events irrespective of their underlying correlation structure and irrespective of the competing risk situation. Moreover, most of these approaches ignore the observational period making the approaches sensitive to changes in the follow-up distribution. Therefore, we generally recommend to apply weighted composite effect measures only when the properties of the method are really understood which usually requires a thorough mathematical look on the methodological aspects.

Two very similar approaches presented by Pocock et al. (2012) and by Buyse (2010) are of special interest here as these methods have been published in high-impact journals in the statistical literature (*Statistics in Medicine*) as well as in the medical literature (*European Heart Journal*). For this reason, it is of strong importance to explore the properties of these approaches in order to justify or to advise against their use. Whereas Pocock et al. (2012) explicitly focus on composite endpoints, Buyse (2010) provides a more general framework for the application of his method. Our comments and recommendations provided below focus only on the specific application for composite endpoints.

G. Rauch et al., *Planning and Analyzing Clinical Trials with Composite Endpoints*, Springer Series in Pharmaceutical Statistics,
https://doi.org/10.1007/978-3-319-73770-6_14

14.1 Proportion in Favor of Treatment and Win Ratio

Effect measures which are based on a priority ranking of the components according to clinical relevance were proposed by Buyse (2010) and Pocock et al. (2012). The idea behind these effect measures is similar: All outcome variables of interest are ordered with respect to their clinical importance. Patients in the intervention group are compared to patients in the control group, and for each comparison the patient with the more favorable outcome with respect to the endpoint of primary priority is determined. If no decision can be made due to censoring or missing values or if both patients are equally good, the comparison is made based on the endpoint of secondary priority and so on. By this, the effect measure gives higher weight to more important endpoints which is meant to reduce masking effects and to improve interpretation of results. The effect measure proposed by Buyse (2010) is referred to as the "proportion in favor of treatment," whereas the effect measure by Pocock et al. (2012) is denoted as the "win ratio." Both approaches are very similar. However, while the proportion in favor of treatment is based on pairwise comparisons between every patient of the intervention group and every patient in the control group and is therefore applicable in the general setting of a randomized controlled trial, the win ratio requires a predefined 1:1 matching from patients in the intervention to the control group. Hence, the approach of Buyse (2010) can be applied in a more general setting, and we will therefore concentrate our considerations on this procedure in the following. However, conclusions drawn for the proportion in favor of treatment are similarly valid for the win ratio.

In the following, let $\mathbf{Z}_{i\cdot}^{I} := \left(Z_{i1}^{I}, \ldots, Z_{ik}^{I}\right), i = 1, \ldots, n_I$, be the identically and independently distributed vectors of random variables with Z_{il}^{I} denoting the outcome of patient i in component EP_l, $l = 1, \ldots, k$, in the intervention group and $\mathbf{Z}_{j\cdot}^{C}$, $j = 1, \ldots, n_C$, is defined analogously for the control group. This notation does not indicate whether the components are time-to-event or binary variables but is used to introduce the general construction method of the approach by Buyse (2010). Define a favor function $\varphi : \mathbb{R}^{2k} \to \{-1, 0, 1\}$ for the comparison of the outcome vectors of two individuals i and j which is given as

$$\varphi\left(\mathbf{Z}_{i\cdot}^{I}, \mathbf{Z}_{j\cdot}^{C}\right) := \begin{cases} 1, & \text{if } \exists\, l \in \{1, \ldots, k\} : \Big(Z_{il}^{I} \text{ is more favorable than } Z_{jl}^{C} \\ & \text{and } \forall h < l : \text{ the comparison between } Z_{il}^{I} \text{ and } Z_{jl}^{C} \text{ is non-informative}\Big), \\ -1, & \text{if } \exists\, l \in \{1, \ldots, k\} : \Big(Z_{il}^{I} \text{ is less favorable than } Z_{jl}^{C} \\ & \text{and } \forall h < l : \text{ the comparison between } Z_{il}^{I} \text{ and } Z_{jl}^{C} \text{ is non-informative}\Big), \\ 0, & \text{if } \forall l \in \{1, \ldots, k\} : \text{ the comparison between } Z_{il}^{I} \text{ and } Z_{jl}^{C} \text{ is non-informative}. \end{cases}$$

The specific definitions for binary and time-to-event variables are given in Rauch et al. (2014).

Using this favor function, the proportion in favor of treatment is defined as

$$\Omega := E\left(\varphi(\mathbf{Z}^{I}_{..}, \mathbf{Z}^{C}_{..})\right).$$

The range of Ω is given by $[-1; 1]$, where positive values favor the intervention.

Rauch et al. (2014) showed that under certain assumptions specified in detail in their work, the proportion in favor of treatment can be expressed as a weighted sum of standard unweighted effect measures for the single components similarly as presented in Chaps. 12 and 13. For components given as binary event indicators and for a constant follow-up time f, the expression Ω can then be written as

$$\Omega = \sum_{l=1}^{k} \left(p^{C}_{EP_l} - p^{I}_{EP_l}\right) \cdot \vartheta^{binary}_{l}(f). \tag{14.1}$$

For multiple time-to-event outcomes, it holds that

$$\Omega = \sum_{l=1}^{k} \frac{\tilde{\theta}_{EP_l} - 1}{\tilde{\theta}_{EP_l} + 1} \cdot \vartheta^{times}_{l}(f), \tag{14.2}$$

where $\vartheta^{*}_{l}, l = 1, \ldots, k$, denote the weights which depend on the fixed follow-up time f and $\tilde{\theta}_{EP_l}$ are the hazard ratios of the individual component hazards which differ from the cause-specific hazards introduced in Sect. 5.2 by ignoring events of other type. The individual hazards should usually not be applied in practice as they ignore the competing risk setting. The representation (14.2) might thus be criticized as being unrealistic in practice. However, Rauch et al. (2014) showed that even for these very simplistic assumptions, the properties of Ω are not adequate, and thus a more complicated parametrization under less restrictive assumptions will even worsen this problem.

14.1.1 Considerations on the Weights

Rauch et al. (2014) showed that the weights $\vartheta^{*}_{l}, l = 1, \ldots, k$, are determined by the follow-up duration, the underlying effects in the components, and (for time-to-event endpoints) by the censoring distribution and the competing risk scenario; compare also Bebu and Lachin (2016). It is thus not guaranteed that the higher weights are really assigned to the more important components. It is important to note that the weights $\vartheta^{*}_{l}, l = 1, \ldots, k$, do not correspond to "free" study parameters which can be chosen in the planning stage. Therefore, it is strictly speaking not correct to refer $\vartheta^{*}_{l}, l = 1, \ldots, k$, as "weights" which is why we omitted the standard notation of weights w_l, $l = 1, \ldots, k$, used in the previous chapters.

In general, the magnitudes of $\vartheta_l^*, l = 1, \ldots, k$, are difficult to predict and sensitively depend on the follow-up duration (Rauch et al. 2014). This makes a correct interpretation of Ω difficult. Another important drawback of the proportion in favor of treatment is that the sum of the "weights" $\vartheta_l^*, l = 1, \ldots, k$, is not standardized, that is, the weights do not sum up to 1. Even worse, the weight sum depends on the study-specific follow-up duration. This makes it difficult to compare Ω between different studies, for example, in the context of a meta-analysis.

14.1.2 Test Problem

The one-sided test problem in terms of the proportion in favor of treatment can be formulated as

$$H_0^{CE} : \Omega \leq 0 \quad \text{versus} \quad H_1^{CE} : \Omega > 0. \tag{14.3}$$

14.1.3 Test Statistic

Buyse (2010) proposed to use a permutation test to test the above hypotheses (14.3). An estimator for the proportion in favor of treatment $\widehat{\Omega}$ is obtained by plugging in the corresponding relative frequency

$$\widehat{\Omega} = \frac{1}{n_I \cdot n_C} \sum_{i=1}^{n_I} \sum_{j=1}^{n_C} \gamma \left(\mathbf{Z}_{i\cdot}^I, \mathbf{Z}_{j\cdot}^C \right).$$

As the calculation of the proportion in favor of treatment requires the comparison of every patient in the intervention group to every patient in the control group and as the permutation test requires to repeat this step many times to receive adequate approximations, the computational effort is high.

14.1.4 Sample Size Calculation

As stated above, the magnitude of the weights $\vartheta_l^*, l = 1, \ldots, k$, is difficult to predict, and therefore valid assumptions on the value of Ω in the planning stage are usually impossible. The sample size can thus only be approximated by simulating reasonable scenarios. However, it is difficult to predict the robustness of such simulations as the behavior and the interaction of the weights are complex. This is a further argument against the application of the method.

14.1.5 Recommendations

To summarize, the magnitude of Ω is difficult to predict and sensitively depends on a number of parameters such as the follow-up duration (Bebu and Lachin 2016; Rauch et al. 2014). To illustrate this problem, Fig. 14.1 shows the outcome of a patient from the intervention group I which is compared to the outcome of a patient from the control group C. In situation a), the patient in the intervention group wins over the patient from the control with respect to the endpoint of primary importance "death." In situation b), the censoring time was reduced. Now, the event "death" is no longer observed, and thus the patient from the control group wins over the patient from the intervention group based on the endpoint of secondary importance "myocardial infarction." As a consequence, a valid interpretation of Ω is challenging.

Another problem related to the proportion in favor of treatment is that the determination of the pairwise "winner" does not result in a well-defined ordering. Figure 14.2 shows a situation comparing the outcomes of three patients (A, B, C). Patient A wins over patient B, as A lives longer than B. Patient B wins over patient C, as the comparison with respect to "death" is uninformative, but patient B experiences the event "myocardial infarction" at a later time point than patient C. So, A is better than B which is better than C. However, comparing patients A and C directly, patient

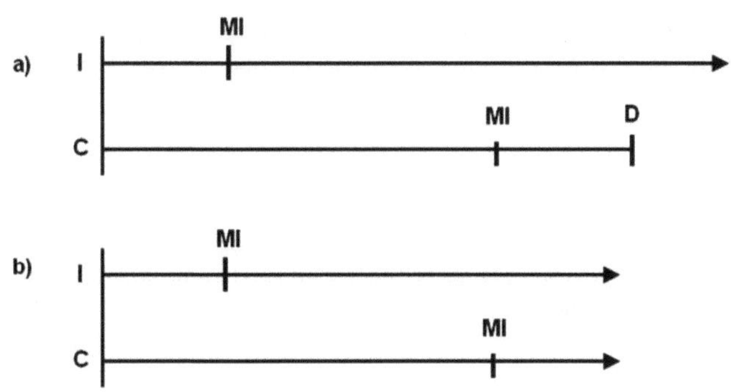

Fig. 14.1 Influence of censoring on the proportion in favor of treatment for two components, where death (D) has first priority and myocardial infarction (MI) has second priority

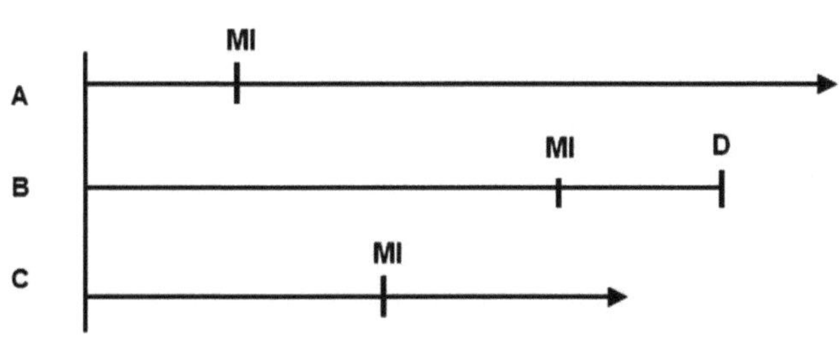

Fig. 14.2 Pairwise comparison of three patient outcomes, where death (D) has first priority and myocardial infarction (MI) has second priority

C wins over A as the comparison with respect to "death" is uninformative, but patient C experiences the event "myocardial infarction" at a later time point than patient C.

For the above reasons, we do not recommend to use the proportion in favor of treatment as a weighted effect measure in the context of composite endpoints. However, we still provide an R code to calculate this effect measure and to perform the corresponding permutation test in Chap. 15 of this book in order to allow the user to make his own experiences with the method. Note that the win ratio and the proportion in favor of treatment can also be used as effect measures for a single outcome variable. Our recommendations and doubts presented in here only refer to the application on a composite endpoint with priority ranked components.

References

Bakal, J. A., Westerhout, C. M., & Armstrong, P. W. (2015). Impact of weighted composite compared to traditional composite endpoints for the design of randomized controlled trials. *Statistical Methods in Medical Research, 24*, 980–988.

Bakal, J. A., Westerhout, C. M., Cantor, W. J., Fernández-Avilés, F., Welch, R. C., Fitchett, D., et al. (2013). Evaluation of early percutaneous coronary intervention vs. standard therapy after fibrinolysis for ST-segment elevation myocardial infarction: Contribution of weighting the composite endpoint. *European Heart Journal, 34*, 903–908.

Bebu, I., & Lachin, J. M. (2016). Large sample inference for a win ratio analysis of a composite outcome based on prioritized outcomes. *Biostatistics, 17*, 178–187.

Buyse, M. (2010). Generalized pairwise comparisons of prioritized outcomes in the two-sample problem. *Statistics in Medicine, 29*, 3245–3257.

Hallstrom, A. P., Litwin, P. E., & Weaver, W. D. (1992). A method of assigning scores to the components of a composite outcome: An example from the MITI trial. *Controlled Clinical Trials, 13*, 148–155.

Lachin, J. M., & Bebu, I. (2015). Application of the Wei-Lachin multivariate one-directional test to multiple event-time outcomes. *Clinical Trials, 12*, 627–633.

Pocock, S. J., Ariti, C. A., Collier, T. J., & Wang, D. (2012). The win ratio: A new approach to the analysis of composite endpoints in clinical trials based on clinical priorities. *European Heart Journal, 33*, 176–182.

Rauch, G., Jahn-Eimermacher, A., Brannath, W., & Kieser, M. (2014). Opportunities and challenges of combined effect measures based on prioritized outcomes. *Statistics in Medicine, 33*, 1104–1120.

Chapter 15
Related Software Code

This chapter provides exemplary source code written in the software R implementing different approaches outlined in Part IV of this book.

15.1 Normal Approximation Test for Weighted Risk Difference

```
library(compiler)

##################################################
# Main function to calculate the point estimator #
# and the normal approximation test for the      #
# weighted risk difference                       #
# based on three disjoint components             #
##################################################


##################################################
# Parameters to specify                          #
##################################################
#
#
# Before you start the program it is necessary to
# import your data into R.
#
# Required structure of the data set "data":
```


G. Rauch et al., *Planning and Analyzing Clinical Trials with Composite Endpoints*, Springer Series in Pharmaceutical Statistics,
https://doi.org/10.1007/978-3-319-73770-6_15

```
# 1.column (id)       = Patient ID
# 3.column (X1)       = Binary event indicator for
#                       ep2 (Component 1)
# 4.column (X2)       = Binary event indicator for
#                       EP2 (Component 2)
# 5.column (X3)       = Binary event indicator for
#                       EP3 (Component 3)
# 6.column (XCE)      = Binary event indicator for
#                       CE (Composite)
# 7.column (group)    = Group allocation
#-----------------------------------------------------
# Other required parameters
#-----------------------------------------------------
# w_ep2               = Weight for ep2
# w_ep2               = Weight for EP2
# w_ep3               = Weight for EP3
#####################################################

weightedRD<-function(w_ep1,w_ep2,w_ep3,data){


#-----------------------------------------------#
# Calculate risk difference and p-value         #
# of the normal approximation test              #
# as a reference                                #
#-----------------------------------------------#

help<-split(data,data$group)
dataI<-data.frame(help[2])
dataC<-data.frame(help[1])

pI<-mean(dataI$X1.XCE)
pC<-mean(dataC$X0.XCE)
nI<-length(dataI$X1.XCE)
nC<-length(dataC$X0.XCE)
p<-(pI*nI+pC*nC)/(nI+nC)
RD_ce<-pC-pI
T_ce<-RD_ce/sqrt(p*(1-p)*(nI+nC)/(nI*nC))
P_ce_onesided<-round(1-pnorm(T_ce),6)

#-----------------------------------------------#
# Calculate risk difference and p-value         #
# of the normal approximation test              #
```

```
# for the components                              #
# as a reference                                  #
#-------------------------------------------------#

pI_ep1<-mean(dataI$X1.X1)
pC_ep1<-mean(dataC$X0.X1)
p_ep1<-(pI_ep1*nI+pC_ep1*nC)/(nI+nC)
RD_ep1<-pC_ep1-pI_ep1
T_ep1<-RD_ep1/sqrt(p_ep1*(1-p_ep1)*(nI+nC)/(nI*nC))
P_ep1_onesided<-round(1-pnorm(T_ep1),6)

#--------------------------------------
pI_ep2<-mean(dataI$X1.X2)
pC_ep2<-mean(dataC$X0.X2)
p_ep2<-(pI_ep2*nI+pC_ep2*nC)/(nI+nC)
RD_ep2<-pC_ep2-pI_ep2
T_ep2<-RD_ep2/sqrt(p_ep2*(1-p_ep2)*(nI+nC)/(nI*nC))
P_ep2_onesided<-round(1-pnorm(T_ep2),6)

#--------------------------------------
pI_ep3<-mean(dataI$X1.X3)
pC_ep3<-mean(dataC$X0.X3)
p_ep3<-(pI_ep3*nI+pC_ep3*nC)/(nI+nC)
RD_ep3<-pC_ep3-pI_ep3
T_ep3<-RD_ep3/sqrt(p_ep3*(1-p_ep3)*(nI+nC)/(nI*nC))
P_ep3_onesided<-round(1-pnorm(T_ep3),6)

#-------------------------------------------------#
# Calculate weighted risk difference and          #
# p-value of the normal approximation test        #
#-------------------------------------------------#

pw_I<-w_ep1*pI_ep1+w_ep2*pI_ep2+w_ep3*pI_ep3
pw_C<-w_ep1*pC_ep1+w_ep2*pC_ep2+w_ep3*pC_ep3
pw<-(pw_I*nI+pw_C*nC)/(nI+nC)
RD_weight<-pw_C-pw_I
T_weight<-RD_weight/sqrt(pw*(1-pw)*(nI+nC)/(nI*nC))
P_weight_onesided<-1-pnorm(T_weight)

out<-matrix(c(RD_ce, RD_ep1, RD_ep2, RD_ep3,
              RD_weight,
            P_ce_onesided, P_ep1_onesided,
            P_ep2_onesided, P_ep3_onesided,
            P_weight_onesided),
            nrow=5,ncol=2,byrow=FALSE)
```

```
dimnames(out) = list(c("Risk Difference CE",
 "Risk Difference EP1",
 "Risk Difference EP2",
 "Risk Difference EP3",
 "Weighted Risk Difference CE"),
c("Estimator",
"p-value (Normal Approximation Test)"))

print("Weights for EP1,EP2,EP3")
print(c(w_ep1,w_ep2,w_ep3))
return(out)

}

weightedRD <- cmpfun(weightedRD)

#**************************************#
# Exemplary calls                      #
#**************************************#

set.seed(170)

#--------------------------------------#
# Generate data where each event in    #
# one of the components                #
# also refers to an event in the       #
# composite                            #
#--------------------------------------#

#--------------------------------------#
# Generate event indicators for three #
# components                           #
#--------------------------------------#

statusI<-rmultinom(100, 1,
                   c(0.1,0.2,0.3,0.4))
statusC<-rmultinom(150, 1,
                   c(0.3,0.3,0.3,0.1))


X1<-c(statusI[1,], statusC[1,])
X2<-c(statusI[2,], statusC[2,])
X3<-c(statusI[3,], statusC[3,])
```

```
#--------------------------------------#
# Composite equals 1 whenever at       #
# least one of the components is 1     #
#--------------------------------------#

XCE<-rep(0,250)
XCE[X1==1 | X2==1 | X3==1] <-1

data<-data.frame(id=c(1:250),
                 X1=X1,
                 X2=X2,
                 X3=X3,
                 XCE=XCE,
                 group=c(rep(1,100),
                         rep(0,150)))
print(data)

weightedRD(0.5,0.4,0.1,data)
weightedRD(0.1,0.4,0.5,data)
weightedRD(1,1,1,data)
```

15.2 Permutation Test for Weighted All-Cause Hazard Ratio

```
library(compiler)
library(survival)
library(parallel)

##################################################
# Main function to calculate the point estimator #
# and the permutation test for the               #
# weighted all-cause hazard ratio                #
# based on three components                      #
##################################################

##################################################
# Parameters to specify                          #
##################################################
#
#
# Before you start the program it is necessary to
# import your data into R.
```

```
#
# Required structure of the data set "data":
#
# 1.column (id)       = Patient ID
# 2.column (times)    = Event time
# 3.column (status1)  = Binary event indicator for
#                       EP1 (Component 1)
# 4.column (status2)  = Binary event indicator for
#                       EP2 (Component 2)
# 5.column (status3)  = Binary event indicator for
#                       EP3 (Component 3)
# 6.column (statusCE)= Binary event indicator for
#                       CE (Composite)
# 7.column (group)    = Group allocation
#---------------------------------------------------
# Other required parameters
#---------------------------------------------------
# w_ep1               = Weight for EP1
# w_ep2               = Weight for EP2
# w_ep3               = Weight for EP3
###################################################

weightedHR<-function(w_ep1,w_ep2,w_ep3,data){


#-----------------------------------------------#
# Calculate hazard ratio and p-value            #
# from the Cox-Model for the composite          #
# endpoint as a reference                       #
#-----------------------------------------------#

cox_ce<-coxph(Surv(times,statusCE)~group,data)
Waldtest_ce<--(cox_ce$coefficients)/
(sqrt(cox_ce$var))
P_ce_onesided<-round(1-pnorm(Waldtest_ce),6)
HR_ce<-1/exp(cox_ce$coefficients)

#-----------------------------------------------#
# Calculate hazard ratio and p-value            #
# from the Cox-Model for the components         #
# as a reference                                #
#-----------------------------------------------#
```

```
cox_ep1<-coxph(Surv(times,status1)~group,data)
Waldtest_ep1<--(cox_ep1$coefficients)/
               (sqrt(cox_ep1$var))
P_ep1_onesided<-round(1-pnorm(Waldtest_ep1),6)
HR_ep1<-1/exp(cox_ep1$coefficients)

cox_ep2<-coxph(Surv(times,status2)~group,data)
Waldtest_ep2<--(cox_ep2$coefficients)/
               (sqrt(cox_ep2$var))
P_ep2_onesided<-round(1-pnorm(Waldtest_ep2),6)
HR_ep2<-1/exp(cox_ep2$coefficients)

cox_ep3<-coxph(Surv(times,status3)~group,data)
Waldtest_ep3<--(cox_ep3$coefficients)/
               (sqrt(cox_ep3$var))
P_ep3_onesided<-round(1-pnorm(Waldtest_ep3),6)
HR_ep3<-1/exp(cox_ep3$coefficients)

#-----------------------------------------------#
# Calculate hazard estimators                   #
# from the Weibull-Model                        #
#-----------------------------------------------#


weibull_ep1<-survreg(Surv(times,status1)~group,
                     data, dist='weibull')
H_I_ep1<-exp(-weibull_ep1$coefficients[1]-
               weibull_ep1$coefficients[2])
H_C_ep1<-exp(-weibull_ep1$coefficients[1])


weibull_ep2<-survreg(Surv(times,status2)~group,
                     data, dist='weibull')
H_I_ep2<-exp(-weibull_ep2$coefficients[1]-
               weibull_ep2$coefficients[2])
H_C_ep2<-exp(-weibull_ep2$coefficients[1])


weibull_ep3<-survreg(Surv(times,status3)~group,
                     data, dist='weibull')
H_I_ep3<-exp(-weibull_ep3$coefficients[1]-
               weibull_ep3$coefficients[2])
H_C_ep3<-exp(-weibull_ep3$coefficients[1])
```

```
#--------------------------------------------#
# Calculate weighted all-cause hazard ratio   #
#--------------------------------------------#

HR_weight<-(w_ep1*H_C_ep1
            + w_ep2*H_C_ep2+ w_ep3*H_C_ep3)/
  (w_ep1*H_I_ep1 + w_ep2*H_I_ep2+ w_ep3*H_I_ep3)

#--------------------------------------------#
# Calculate p-value based on permutation test #
# based on 1000 permutations                  #
#--------------------------------------------#

HRdist_weight<-c()


for (i in 1:1000){
  data$group<-sample(data$group, replace=FALSE)

  weibull_ep1<-survreg(Surv(times,status1)~group,
                       data, dist='weibull')
  H_I_ep1<-exp(-weibull_ep1$coefficients[1]-
                weibull_ep1$coefficients[2])
  H_C_ep1<-exp(-weibull_ep1$coefficients[1])

  weibull_ep2<-survreg(Surv(times,status2)~group,
                       data, dist='weibull')
  H_I_ep2<-exp(-weibull_ep2$coefficients[1]-
                weibull_ep2$coefficients[2])
  H_C_ep2<-exp(-weibull_ep2$coefficients[1])

  weibull_ep3<-survreg(Surv(times,status3)~group,
                       data, dist='weibull')
  H_I_ep3<-exp(-weibull_ep3$coefficients[1]-
                weibull_ep3$coefficients[2])
  H_C_ep3<-exp(-weibull_ep3$coefficients[1])

  HRdist_weight[i]<-(w_ep1*H_C_ep1
              + w_ep2*H_C_ep2+ w_ep3*H_C_ep3)/
    (w_ep1*H_I_ep1 + w_ep2*H_I_ep2+ w_ep3*H_I_ep3)


}
```

```
P_weight_onesided_perm<-round(1-ecdf(HRdist_weight)
                                (HR_weight))

out<-matrix(c(HR_ce,HR_ep1,HR_ep2,HR_ep3,HR_weight,
            P_ce_onesided, P_ep1_onesided,
            P_ep2_onesided,P_ep3_onesided,
            P_weight_onesided_perm),
            nrow=5,ncol=2,byrow=FALSE)
dimnames(out) = list(c("CE Cox","EP1 Cox","EP2 Cox",
                "EP3 Cox","WeightedCE"),
                c("HR","p-value (Cox/Permutation)"))

print("Weights for EP1,EP2,EP3")
print(c(w_ep1,w_ep2,w_ep3))
return(out)

}

weightedHR <- cmpfun(weightedHR)



#**************************************#
# Exemplary calls                      #
#**************************************#
set.seed(200)
#--------------------------------------#
# Generate data where each event in    #
# the first endpoint (main component) #
# also refers to an event in the       #
# second component (composite).        #
#--------------------------------------#

#--------------------------------------#
# Specify all-cause hazards            #
#--------------------------------------#

haz_I<-0.5
haz_C<-0.8

#--------------------------------------#
# Generate event indicators for two    #
# components                           #
#--------------------------------------#
```

```
statusI<-rmultinom(200, 1,
        c(0.15/haz_I,0.1/haz_I,0.05/haz_I,
        1-(0.15+0.1+0.05)/haz_I))
statusC<-rmultinom(200, 1,
        c(0.2/haz_C,0.2/haz_C,0.2/haz_C,
        1-(0.2+0.2+0.2)/haz_C))


status1<-c(statusI[1,], statusC[1,])
status2<-c(statusI[2,], statusC[2,])
status3<-c(statusI[3,], statusC[3,])

#-------------------------------------#
# Composite equals 1 whenever at      #
# least one of the components is 1    #
#-------------------------------------#
statusCE<-rep(0,400)
statusCE[status1==1|status2==1|status3==1]<-1


data<-data.frame(id=c(1:400),
            times=c(-log(runif(200,0,1))/(haz_I),
                 -log(runif(200,0,1))/(haz_C)),
                 status1=status1,
                 status2=status2,
                 status3=status3,
                 statusCE=statusCE,

                 group=c(rep(1,200),rep(0,200)))

print("True Hazard Ratios CE,EP1,EP2,EP3")
print(c(0.8/0.5,0.2/0.15,0.2/0.1,0.2/0.05))


print("Estimated Hazard Ratios and p-values")
weightedHR(1,1,1,data)
weightedHR(0.5,0.4,0.1,data)
weightedHR(0.1,0.4,0.5,data)
```

15.3 Permutation Test for Proportion in Favor of Treatment

15.3.1 Proportion in Favor of Treatment for Binary Endpoints

```
###################################################
# Main function to calculate the point estimator, #
# confidence interval                             #
# and permutation test for the                    #
# proportion in favor of treatment                #
# based on two binary endpoints                   #
###################################################

###################################################
# Parameters to specify                           #
###################################################
#
#
# Before you start the program it is necessary to
# import your data into R.
#
# Required structure of the data set "data":
#
# 1.column (id)     = Patient ID
# 2.column (Xprim)  = Binary event indicator for
#                     the most important component
# 3.column (Xsek)   = Binary event indicator for
#                     the second important component
# 4.column (group)  = group allocation
#--------------------------------------------------
# Other required parameters
#--------------------------------------------------
# nsim          = number of data sets simulations
# alpha         = significance level to compute
#                 two-sided 1-alpha confidence
#                 interval
###################################################

propfavor<-function(data,nsim,alpha){

#-------------------------------------------#
# Prepare data set for further calculations  #
#-------------------------------------------#
```

```
t<-split(data,data$group)
data_original_I<-data.frame(t[2])
data_original_I<-data.frame(
                  id=data_original_I$X1.id,
                  Xprim_I=data_original_I$X1.Xprim,
                  Xsek_I=data_original_I$X1.Xsek)
data_original_C<-data.frame(t[1])
data_original_C<-data.frame(
                  id=data_original_C$X0.id,
                  Xprim_C=data_original_C$X0.Xprim,
                  Xsek_C=data_original_C$X0.Xsek)


#----------------------------------------------#
# Define three help functions "controlmatric",#
# "compare" and "pointestimator"              #
#----------------------------------------------#



#----------------------------------------------#
# "controlmatric" = Function to generate all  #
# permutations of the control group           #
#----------------------------------------------#

controlmatrix<-function(outcome_C){
  n<-dim(outcome_C)[1]
  helpmatrix<-matrix(ncol=2*n,nrow=n)
  helpmatrix<-as.data.frame(helpmatrix)
  for (j in 1:n){
    m1<-outcome_C[1:j,-1]
    m2<-outcome_C[-(1:j),-1]
    l<-rbind(m2,m1)
    k<-2*j-1
    helpmatrix[,k]<-l[,1]
    helpmatrix[,k+1]<-l[,2]
    names(helpmatrix)[k]<-paste("Xprim_C",
                                j,sep="")
    names(helpmatrix)[k+1]<-paste("Xsek_C",
                                  j,sep="")
  }
  return(helpmatrix)
}
```

```
#---------------------------------------------#
# "compare" :Function for all pairwise        #
# comparisons between two groups              #
#---------------------------------------------#

compare<-function(outcome_I,outcomematrix){
  n1<-dim(outcome_I)[1]
  n2<-dim(outcomematrix)[1]
  indicator2<-matrix(ncol=n2,nrow=n1)
  indicator<-matrix(ncol=n2,nrow=n1)

  #---------------------------------------------#
  # comparison with respect to second important#
  # important component                         #
  #---------------------------------------------#

  for(i in 1:n1){
    for(j in 1:n2){
      if(outcome_I$Xsek_I[i]==0
          & outcomematrix[i,2*j]==1){
        indicator2[i,j]=1
      } else if(outcome_I$Xsek_I[i]==1
                & outcomematrix[i,2*j]==0){
        indicator2[i,j]=-1
      } else {
        indicator2[i,j]=0
      }
    }
  }

  #----------------------------------------------#
  # comparison with respect to most important    #
  # component                                    #
  #----------------------------------------------#

  for(i in 1:n1){
    for(j in 1:n2){
      #no missing in both groups
      if((is.na(outcome_I$Xprim_I[i]) == FALSE)
    &(is.na(outcomematrix[i,2*j-1]) == FALSE)){
        if(outcome_I$Xprim_I[i]==0
            & outcomematrix[i,2*j-1]==1){
          indicator[i,j]=1
        } else if(outcome_I$Xprim_I[i]==1
                & outcomematrix[i,2*j-1]==0) {
```

```
          indicator[i,j]=-1
        } else {
          indicator[i,j]=indicator2[i,j]
        }
        #no missing in control group
      } else if((is.na(outcome_I$Xprim_I[i])==TRUE)
      & (is.na(outcomematrix[i,2*j-1]) == FALSE)){
        if(outcomematrix[i,2*j-1]==1){
          indicator[i,j]=1
        } else {
          indicator[i,j]=indicator2[i,j]
        }
        #no missing in intervention group
      } else if((is.na(outcome_I$Xprim_I[i])==FALSE)
          (is.na(outcomematrix[i,2*j-1]) == TRUE)){
        if(outcome_I$Xprim_I[i]==1){
          indicator[i,j]=indicator2[i,j]
        } else {
          indicator[i,j]=-1
        }
     #missing in control group and intervention group
      } else {
        indicator[i,j]= indicator2[i,j]
      }
    }
  }
  return(indicator)
}

#----------------------------------------------#
# "Pointestimator": Function to calculate the #
# point estimator                              #
#----------------------------------------------#

Pointestimator <- function(indicatormatrix){
  pointestimator <- NULL
  n1 <- dim(indicatormatrix)[1]
  for (i in 1:n1){
    pointestimator[i]<-mean(indicatormatrix[i,])
  }
  pointestimator <- mean(pointestimator)
  return(pointestimator)
}
```

```
#------------------------------------------------#
# Calculation of the point estimator             #
#------------------------------------------------#

indicatormatrix <- compare(data_original_I,
                  controlmatrix(data_original_C))

Pointestimator_original <- Pointestimator(
                              indicatormatrix)


#------------------------------------------------#
# Preparation of the data for permutation        #
#------------------------------------------------#

names(data_original_I)[2] <- paste("Xprim")
names(data_original_I)[3] <- paste("Xsek")
names(data_original_C)[2] <- paste("Xprim")
names(data_original_C)[3] <- paste("Xsek")

n1 <- dim(data_original_I)[1]
n2 <- dim(data_original_C)[1]
m <- n1+n2

intervention <- rbind(data_original_C,
                      data_original_I)
intervention$obs <- seq(1,m)

#-----------------------------------------------#
# Permutation                                   #
#-----------------------------------------------#

distribution <- vector(length=nsim)
for(h in 1:nsim){

  rand <- sample(c(rep(1,n1), rep(2,n2)))

  perm <- cbind(rand,intervention)

  outcome_permuted_I <- subset(perm,perm$rand==1)
  names(outcome_permuted_I)[c(3,4)]<-c("Xprim_I",
                                        "Xsek_I")
  outcome_permuted_I[,1] <- seq(1,n1)
  outcome_permuted_I <- outcome_permuted_I[,-2]
```

```
    outcome_permuted_C <- subset(perm,perm$rand==2)
    names(outcome_permuted_C)[c(3,4)]<-c("Xprim_C1",
                                         "Xsek_C1")
    outcome_permuted_C <- outcome_permuted_C[,-2]
    outcome_permuted_C[,1] <- seq(1,n2)

    distribution[h] <-Pointestimator(
                compare(outcome_permuted_I,
                controlmatrix(outcome_permuted_C)))
  }

    #--------------------------------------------#
    # Calculation of the one-sided p-value      #
    #--------------------------------------------#

    Fn <-ecdf(distribution)
    p_value <-round(1-Fn(Pointestimator_original),
                    digits=5)
    #---------------------------------------------#
    # Calculation of the (one-sided) 1-alpha      #
    # confidence interval                         #
    #---------------------------------------------#

    confintlow <-Pointestimator_original-
      qnorm(1-0.5*alpha)*sd(distribution)
    confintup <- Pointestimator_original+
      qnorm(1-0.5*alpha)*sd(distribution)
    #--------------------------------------------#
    # Histogram of the null distribution of the #
    # proportion in favor of treatment          #
    #--------------------------------------------#

    hist(distribution, col="gray", freq=FALSE,
    main="Null distr. of the Prop. in Favor of Treat.",
         xlim=c(-1,1))
    abline(v=Pointestimator_original,col="blue")
    abline(v=confintup,col="red")
    abline(v=confintlow,col="red")
    legend(-1,1.5,c("Point Estimator",
                    "Condfidence Interval"),
           fill=c("blue", "red"))

    res <- data.frame(Pointestimator_original,
                      sd(distribution),
                      confintlow,
```

```
                         confintup,p_value)
  names(res)[1]<-
paste("Estimated Prop.in Favor of Treat.")
  names(res)[2]<-paste("Estimated STD (under H0)")
  names(res)[3]<-
paste("Lower 1-alpha Confidence Limit")
  names(res)[4]<-
paste("Upper 1-alpha Confidence Limit")
  names(res)[5]<-paste("One-Sided p-Value")
  print(res)
}

#*************************************#
# Exemplary calls                     #
#*************************************#

#-------------------------------------#
# Generate exemplary data set         #
#-------------------------------------#

data<- data.frame(id=1:40,
Xprim=c(rbinom(20,1,0.4),rbinom(20,1,0.5)),
Xsek=c(rbinom(20,1,0.2),rbinom(20,1,0.3)),
group=as.factor(c(rep(1,20),rep(0,20))))

example <- propfavor(data,1000,0.05)

#################################################
# Remark:                                       #
# The program takes about 1.76 days             #
# for nsim= 1000 simulations with 500 patient   #
# data in each group (control and intervention).#
# Processor: AMD Athlon (tm) Dual Core          #
# Processor 5400B 2.80Ghz                       #
#################################################
```

15.3.2 *Proportion in Favor of Treatment for Time-to-Event Endpoints*

```
###################################################
# Main function to calculate the point estimator, #
# confidence interval                             #
# and permutation test for the                    #
```

```
# proportion in favor of treatment                  #
# based on two time-to-event endpoints              #
#####################################################

#####################################################
# Parameters to specify                             #
#####################################################
#
#
# Before you start the program it is necessary to
# import your data into R.
#
# Required structure of the data set "times":
#
# 1.column (id)    = Patient ID
# 2.column (Tprim) = survival time of the most
#                    important component;
#                    include missing if censored;
#                    do not report the censoring
#                    time here!
# 3.column (Tsek)  = survival time of the second
#                    important component;
#                    include missing if censored;
#                    do not report the censoring
#                    time here
# 4.column (Tcens) = individual censoring time
# 5.column (group) = group allocation
#----------------------------------------------------
# Other required parameters
#----------------------------------------------------
# nsim             = number of data sets simulations
# alpha            = significance level to compute
#                    two-sided 1-alpha confidence
#                    interval
#####################################################

propfavor<-function(times,nsim,alpha){


  #-----------------------------------------------#
  # Prepare data set for further calculations   #
  #-----------------------------------------------#

test1<-rep(0,length(times))
```

```
test1[is.na(times$Tprim)==FALSE &&
        times$Tprim>times$Tcens]<-1

test2<-rep(0,length(times))
test2[is.na(times$Tsek)==FALSE &&
        times$Tsek>times$Tcens]<-1

if(sum(test1)>0 | sum(test2)>0){
print("Event times (Trim,Tsek) must not include"
,"censoring times.")
print("Censoring times are captured by",
"the variable Tcens.")
print("If the event time is censored,",
"set Tprim and/or Tsek to NA.")

}else{


t<-split(times,times$group)
times_original_I<-data.frame(t[2])
times_original_I<-data.frame(
id=times_original_I$X1.id,
                Tprim_I=times_original_I$X1.Tprim,
                Tsek_I=times_original_I$X1.Tsek,
                Tcens_I=times_original_I$X1.Tcens)
times_original_C<-data.frame(t[1])
times_original_C<-data.frame(
id=times_original_C$X0.id,
                Tprim_C=times_original_C$X0.Tprim,
                Tsek_C=times_original_C$X0.Tsek,
                Tcens_C=times_original_C$X0.Tcens)


#---------------------------------------------#
# Define three help functions "controlmatric",#
# "compare" and "pointestimator"              #
#---------------------------------------------#


#---------------------------------------------#
# "controlmatric" = Function to generate all  #
# permutations of the control group           #
#---------------------------------------------#

controlmatrix <- function(times_C){
```

```
  n <- dim(times_C)[1]
  helpmatrix <-matrix(ncol=3*n,nrow=n)
  helpmatrix <- as.data.frame(helpmatrix)
  for (j in 1:n){
    m1 <- times_C[1:j,-1]
    m2 <- times_C[-(1:j),-1]
    l <- rbind(m2,m1)
    k <- 3*j-2
    helpmatrix[,k] <- l[,1]
    helpmatrix[,k+1] <- l[,2]
    helpmatrix[,k+2] <- l[,3]
    names(helpmatrix)[k]<-paste("Tprim_C",
j,sep="")
    names(helpmatrix)[k+1]<-paste("Tsek_C",
j,sep="")
    names(helpmatrix)[k+2]<-paste("Tcens_C",
j,sep="")
  }
  return(helpmatrix)
}

#-----------------------------------------------#
# "compare" :Function for all pairwise          #
# comparisons between two groups                #
#-----------------------------------------------#

compare <- function(times_I,times_Cmatrix){
  n1 <- dim(times_I)[1]
  n2 <- dim(times_Cmatrix)[1]
  indicator2 <- matrix(ncol=n2,nrow=n1)
  indicator <- matrix(ncol=n2,nrow=n1)

  #-----------------------------------------------#
  # comparison with respect to second important#
  # important component                         #
  #-----------------------------------------------#

for(i in 1:n1){
for(j in 1:n2){
#no censorship in both groups
if((is.na(times_I$Tsek_I[i]) == FALSE )
```

```
 & (is.na(times_Cmatrix[i,3*j-1]) == FALSE)){
if(times_I$Tsek_I[i]>times_Cmatrix[i,3*j-1]){
indicator2[i,j]=1
} else if( times_I$Tsek_I[i]<
 times_Cmatrix[i,3*j-1]){
indicator2[i,j]=-1
} else {
indicator2[i,j]=0
}
#no censorship in control group
} else if((is.na(times_I$Tsek_I[i]) == TRUE)
& (is.na(times_Cmatrix[i,3*j-1]) == FALSE)){
if(times_I$Tcens_I[i]>times_Cmatrix[i,3*j-1]){
indicator2[i,j]=1
} else {
indicator2[i,j]=0
}
#no censorship in intervention group
} else if((is.na(times_I$Tsek_I[i]) == FALSE)
 & (is.na(times_Cmatrix[i,3*j-1]) == TRUE)){
if(times_I$Tsek_I[i]>=times_Cmatrix[i,3*j]){
indicator2[i,j]=0
} else {
indicator2[i,j]=-1
}
#censorship in both groups
} else {
indicator2[i,j]=0
}
}
}

#-----------------------------------------------#
# comparison with respect to most important     #
# component                                     #
#-----------------------------------------------#

for(i in 1:n1){
for(j in 1:n2){
#no censorship in both groups
if((is.na(times_I$Tprim_I[i]) == FALSE)
 &(is.na(times_Cmatrix[i,3*j-2]) == FALSE)){
if(times_I$Tprim_I[i]>times_Cmatrix[i,3*j-2]){
indicator[i,j]=1
} else if(times_I$Tprim_I[i]
```

```
<times_Cmatrix[i,3*j-2]){
indicator[i,j]=-1
} else {
indicator[i,j]=indicator2[i,j]
}
#no censorship in control group
} else if((is.na(times_I$Tprim_I[i]) == TRUE )
& (is.na(times_Cmatrix[i,3*j-2]) == FALSE)){
if(times_I$Tcens_I[i]>times_Cmatrix[i,3*j-2]){
indicator[i,j]=1
} else {
indicator[i,j]=indicator2[i,j]
}
#no censorship in intervention group
} else if((is.na(times_I$Tprim_I[i]) == FALSE)
 &(is.na(times_Cmatrix[i,3*j-2]) == TRUE)){
if(times_I$Tprim_I[i]>=times_Cmatrix[i,3*j]){
indicator[i,j]=indicator2[i,j]
} else {
indicator[i,j]=-1
}
#censorship in control group
} else {
indicator[i,j]= indicator2[i,j]
}
}
}
return(indicator)
}

#-----------------------------------------------#
# "Pointestimator": Function to calculate the #
# point estimator                               #
#-----------------------------------------------#

Pointestimator <- function(indicatormatrix){
  pointestimator <- NULL
  n1 <- dim(indicatormatrix)[1]
  for (i in 1:n1){
    pointestimator[i]<-mean(indicatormatrix[i,])
  }
  pointestimator <- mean(pointestimator)
  return(pointestimator)
}
```

```
#-----------------------------------------------#
# Calculation of the point estimator            #
#-----------------------------------------------#
indicatormatrix <- compare(times_original_I,
               controlmatrix(times_original_C))


pointestimator <- NULL
n1 <- dim(indicatormatrix)[1]
for (i in 1:n1){
  pointestimator[i] <- mean(indicatormatrix[i,])
}
Pointestimator_original <- mean(pointestimator)

#-----------------------------------------------#
# Preparation of the data for permutation       #
#-----------------------------------------------#

  names(times_original_I)[2] <- paste("Tprim")
  names(times_original_I)[3] <- paste("Tsek")
  names(times_original_I)[4] <- paste("Tcens")
  names(times_original_C)[2] <- paste("Tprim")
  names(times_original_C)[3] <- paste("Tsek")
  names(times_original_C)[4] <- paste("Tcens")

  n1 <- dim(times_original_I)[1]
  n2 <- dim(times_original_C)[1]
  m <- n1+n2

  intervention <-rbind(times_original_C,
                       times_original_I)
  intervention$obs <- seq(1,m)

  #-----------------------------------------------#
  # Permutation                                   #
  #-----------------------------------------------#
  distribution <- vector(length=nsim)
  for(h in 1:nsim){

    rand <- sample(c(rep(1,n1), rep(2,n2)))

    perm <- cbind(rand,intervention)


    times_permuted_I <- subset(perm,perm$rand==1)
```

```
    names(times_permuted_I)[c(3,4,5)] <-
            c("Tprim_I","Tsek_I","Tcens_I")
    times_permuted_I[,1] <- seq(1,n1)
    times_permuted_I <- times_permuted_I[,-2]


    times_permuted_C <- subset(perm,perm$rand==2)
    names(times_permuted_C)[c(3,4,5)] <-
            c("Tprim_C1","Tsek_C1","Tcens_C1")
    times_permuted_C <- times_permuted_C[,-2]
    times_permuted_C[,1] <- seq(1,n2)

    distribution[h]<-Pointestimator(compare(
              times_permuted_I,
              controlmatrix(times_permuted_C)))
  }

  #--------------------------------------------#
  # Calculation of the one-sided p-value       #
  #--------------------------------------------#

  Fn <-ecdf(distribution)
  p_value <-round(1-Fn(Pointestimator_original),
                  digits=5)
  #--------------------------------------------#
  # Calculation of the (one-sided) 1-alpha     #
  # confidence interval                        #
  #--------------------------------------------#

  confintlow <-Pointestimator_original-
              qnorm(1-0.5*alpha)*sd(distribution)
  confintup <- Pointestimator_original+
              qnorm(1-0.5*alpha)*sd(distribution)
#--------------------------------------------#
# Histogram of the null distribution of the #
# proportion in favor of treatment          #
#--------------------------------------------#

hist(distribution, col="gray", freq=FALSE,
main="Null dist. of the Prop.in Favor of Treat.",
xlim=c(-1,1))
abline(v=Pointestimator_original,col="blue")
abline(v=confintup,col="red")
abline(v=confintlow,col="red")
legend(-1,1.5,c("Point Estimator",
```

```
"Confidence Interval"),
  fill=c("blue", "red"))

res <- data.frame(Pointestimator_original,
sd(distribution),
confintlow,confintup,p_value)
names(res)[1]<-
paste("Estimated Prop.in Favor of Treat.")
names(res)[2]<-paste("Estimated STD (under H0)")
names(res)[3]<-
paste("Lower 1-alpha Confidence Limit")
names(res)[4]<-
paste("Upper 1-alpha Confidence Limit")
names(res)[5]<-paste("One-Sided p-Value")
print(res)}
}

#***********************************#
# Exemplary Calls                   #
#***********************************#

#-----------------------------------#
# Generate exemplary data set       #
#-----------------------------------#

times<- data.frame(id=1:40,
Tprim=c(-log(runif(20))/0.4,-log(runif(20))/0.5),
Tsek=c(-log(runif(20))/0.2,-log(runif(20))/0.3),
Tcens=rep(2,40),
group=as.factor(c(rep(1,20),rep(0,20))))
times$Tprim[times$Tcens<times$Tprim]<-NA
times$Tsek[times$Tcens<times$Tsek]<-NA

example <- propfavor(times,1000,0.05)

####################################################
# Remark:                                          #
# The program takes about 1.76 days                #
# for nsim= 1000 simulations with 500 patient      #
# data in each group (control and intervention).   #
# Processor: AMD Athlon (tm) Dual Core             #
# Processor 5400B 2.80Ghz                          #
####################################################
```

```
print(paste("Teletrail" [illegible]))
[illegible]

[illegible]
[illegible]
[illegible]
[illegible]
[illegible]
[illegible]
[illegible]
[illegible]
[illegible]
[illegible]
[illegible]
[illegible]
}

#**************************************#
# Exemplary Call                       #
#**************************************#

#--------------------------------------#
# Generate exemplary data set          #
#--------------------------------------#

times<-data.frame(id=1:40,
 Tprim=c(-log(runif(20))/0.4,-log(runif(20))/0.5),
 Tsek=c(-log(runif(20))/0.2,-log(runif(20))/0.3),
 Tcens=rep(2,40),
 group=as.factor(c(rep(1,20),rep(0,20))))
times$Tprim[times$Tcens<times$Tprim]<-NA
times$Tsek[times$Tcens<times$Tsek]<-NA

example <- propfavor(times,1000,0.05)

########################################
# Remark:                              #
# The program takes about 1,76 days    #
# for nsim=1000 simulations with 500 patient #
# data in each group (control and intervention) #
# Processor: AMD Athlon (tm) Dual Core  #
# Processor 5400B 2.80GHz               #
########################################
```

Part V
Descriptive and Confirmatory Evaluation of the Components

Whenever a composite endpoint is used as primary efficacy variable, the related guidelines on composite endpoints unanimously recommend to additionally evaluate the individual components separately; compare Chap. 3 in Part I of the book. Thereby, it is not explicitly stated whether the analysis of the components should be of descriptive or of confirmatory nature. Descriptive analyses of the components are required to complete the overall picture on the treatment effects and to discuss the relevance of the composite effect. For example, a descriptive analysis of the components may reveal an adverse effect in a relevant component thereby questioning the interpretation of the composite effect. There exist a number of commonly applied approaches to provide descriptive analyses of the components, but not all of them can be recommended. In this context, it is important to note that the individual components of a composite define competing risks which has to be accounted for when calculating and interpreting component-specific treatment effects. In this part of the book, we will provide recommendations for an adequate descriptive analysis of the components.

Despite the usefulness of descriptive analyses, they do not provide confirmatory evidence and give no definite rule how to judge the overall treatment efficacy. However, it is usually not feasible to power the trial such that all components can be assessed in a confirmatory way by means of a multiple test problem. But even in case the study is only powered to detect a treatment effect in the composite (or additionally also in the main component), adequate multiple test strategies can provide reasonable chance to get additional confirmatory evidence on the remaining components without the need to increase the sample size. These approaches will also be presented within this part of the book.

Chapter 16
Descriptive Analysis of the Components

The resulting treatment effect of a composite endpoint alone is often difficult to interpret as the individual components do not necessarily contribute the same amounts to this net effect. As discussed before, current guidelines therefore recommend to always analyze the components of composite endpoints separately. The current well-established practice is to provide descriptive analyses of the components and the composite in addition to the confirmatory analysis of the composite. However, there exist a number of different methods to evaluate the single components in a descriptive manner. In this chapter, the most common approaches for a descriptive analysis of the individual components will be presented and discussed. Moreover, we will deduce recommendations for a meaningful presentation and interpretation of the component-specific results.

16.1 Reporting of Estimated Event Probabilities

The most commonly reported descriptive summary measure for a composite endpoint and its components is the corresponding estimated event probability per treatment arm, equivalently denoted as "event rate." Group comparison is then usually performed by evaluating the absolute differences in the group-specific event probabilities at a specific time point. The concept of an event probability which is a quantity within the interval [0; 1] implies that only one event per patient is counted and thus information from recurrent events is neglected. The short expression "event probability" thereby more exactly refers to the probability to observe an event of any type or an event of type l, $l = 1, \ldots, k$, respectively, up to a certain time point t, $t \geq 0$. The event probability as a function in time is also denoted as the "cumulative incidence function"; compare, for example, Beyersmann et al. (2012) and Pintilie (2006).

Using the notations introduced in Chap. 5.2 in Part II of this book, we will again consider a multistate model with one initial state at the time origin $Z(0) = 0$ and k causes of failure, where k refers to the number of components of the composite. As before, $T := \inf\{t > 0 \,|\, Z(t) \neq 0\}$ defines the time the patient moves out of

G. Rauch et al., *Planning and Analyzing Clinical Trials with Composite Endpoints*, Springer Series in Pharmaceutical Statistics,
https://doi.org/10.1007/978-3-319-73770-6_16

the initial state. To introduce the concept of event probabilities, the indices for the group affiliation will first be omitted for the sake of simplicity. Indices will only be reported when explicitly denoting effect measures for the group comparison. The composite event probability which is given by the probability to observe an event of any type up to time t is then given as

$$P_{CE}(t) := P(T \leq t) = 1 - S_{CE}(t),$$

where $S_{CE}(t) := P(T > t)$ is the event time distribution function corresponding to the composite endpoint which is given by

$$S_{CE}(t) = \int_0^t \lambda_{CE}(u)\, du = \int_0^t \sum_{l=1}^{k} \lambda_{EP_l}(u)\, du, \ t \geq 0;$$

compare Beyersmann et al. (2012). The component event probability, which is given by the probability to experience an event of type l, $l = 1, \ldots, k$, up to time t, is given as

$$\begin{aligned} P_{EP_l}(t) &:= P(T \leq t, Z(T) = l) \\ &= \int_0^t S_{CE}^-(u) \lambda_{EP_l}(u)\, du, \ t \geq 0, \ l = 1, \ldots, k, \end{aligned} \tag{16.1}$$

where $S_{CE}^-(t)$ defines the value of $S_{CE}(t)$ just prior to time t. Note that the cumulative incidence functions for each event type sum up to the all-cause cumulative incidence of the composite

$$P_{CE}(t) = \sum_{l=1}^{k} P_{EP_l}(t), \ t \geq 0. \tag{16.2}$$

Using the above notations, a commonly applied effect measure for the composite endpoint and the components is given by the absolute difference in the group-specific event probabilities

$$\Delta_{CE} := P_{CE}^C(t) - P_{CE}^I(t), \tag{16.3}$$

$$\Delta_l := P_{EP_l}^C(t) - P_{EP_l}^I(t), \ t \geq 0, \ l = 1, \ldots, k, \tag{16.4}$$

where C and I now denote the indices for the group affiliation.

In the following, the task will be to define adequate point estimators and related standard errors for $P_{CE}(t)$ and $P_{EP_l}(t)$, $l = 1, \ldots, k$, for each treatment group. Thereby, recall that the event probability is a time-dependent function. Thus, when providing an estimate for the event probability, the observation time point must be reported. In many medical publications, the time dependence is ignored, and estimated event probabilities are reported without explicitly stating the time of assessment. For a meaningful interpretation of event probabilities, however, the observation time point is a mandatory requirement.

Another important aspect when comparing component-specific event probabilities between different treatment groups is that the magnitude of the event probability for a specific component naturally depends on the cause-specific event probabilities of all competing causes of failure which is easily deduced from (16.2). To illustrate this issue, recall the exemplary situation described in Chap. 2 in Part I of this book. Assume that a novel therapeutic intervention in patients with cardiovascular disease is associated with a 1-year mortality of 0.2 as compared to a death rate of 0.4 in the control group. However, within the same cohort, the rate for nonfatal myocardial infarction is 0.3 in the treatment group but only 0.1 in the control. In this situation, it is intuitive that the lower event rate for nonfatal myocardial infarction in the control group is probably due to the fact that many patients had died before experiencing a myocardial infarction. As a consequence, cause-specific event probabilities should not be compared between groups without simultaneously looking at all competing event probabilities.

There exist several commonly met approaches which intend to estimate event probabilities. However, many of these well-established methods result in strongly biased estimates and, as a consequence, should not be used. In the following, we begin by introducing a correct way to estimate event probabilities and corresponding standard errors for the composite endpoint and its components. Subsequently, other commonly applied approaches are discussed which do not result in meaningful estimators and are thus not recommended. Note that estimated event probabilities should generally be reported along with standard errors or confidence intervals to quantify the precision of the estimates.

16.1.1 Estimation by the Empirical Cumulative Incidence Function

To estimate the composite event probability or, equivalently, the cumulative incidence function related to the composite endpoint, the most common nonparametric approach was proposed by Kaplan and Meier (1958). The Kaplan-Meier estimator relates to the inverse of the empirical cumulative incidence function of the composite endpoint and defines an estimator for the composite event time distribution $S_{CE}(t)$ which is given as

$$\widehat{S}_{CE}(t) = \prod_{t_{(i)} \leq t} \left(1 - \frac{\Delta N_{CE}(t_{(i)})}{Y(t_{(i)})}\right), \; t \geq 0, \; i = 1, \ldots, n, \tag{16.5}$$

where $t_{(1)} \leq t_{(2)} \leq \cdots \leq t_{(n)}$ denotes the observed ordered event or censoring times of the individual patients, $\Delta N_{CE}(t_{(i)})$ refers to the total number of observed events at time $t_{(i)}$, and $Y(t_{(i)})$ denotes the set of patients under risk just prior time $t_{(i)}$. The risk set $Y(t_{(i)})$ is given by all patients who are event-free at time $t_{(i)}$ and not censored before time $t_{(i)}$. The probability to observe an event of any type up to time t can thus be estimated as

$$\widehat{P}_{CE}(t) := 1 - \widehat{S}_{CE}(t). \tag{16.6}$$

Estimators for the point-wise standard errors of the inverse Kaplan-Meier estimator are described, for example, in Pintilie (2006). In there, a formula for calculating confidence intervals is also provided based on a $ln(-ln(\cdot))$ transformation to guarantee that the confidence interval lays within [0; 1].

In order to estimate the component-specific cumulative incidence functions for each event type l, $l = 1, \ldots, k$, recall that $P_{EP_l}(t)$ directly depends on $S_{CE}(t)$; compare (16.1). Therefore, an estimator for $P_{EP_l}(t)$ can be derived by plugging in the Kaplan-Meier estimator $\widehat{S}_{CE}(t)$. For $l = 1, \ldots, k$ the resulting estimators are given as

$$\widehat{P}_{EP_l}(t) := \sum_{t_{(i)} \leq t} \widehat{S}_{CE}(t_{(i-1)}) \cdot \frac{\Delta N_{EP_l}(t_{(i)})}{Y(t_{(i)})}, \; t \geq 0, \; i = 1, \ldots, n, \tag{16.7}$$

where $\Delta N_{EP_l}(t_{(i)})$ refers to the total number of observed type-l events at time $t_{(i)}$ and $Y(t_{(i)})$ again denotes the set of patients under risk just prior time $t_{(i)}$. It is easy to verify that

$$\widehat{P}_{CE}(t) = \sum_{l=1}^{k} \widehat{P}_{EP_l}(t),$$

which defines an intuitive relation between the estimators for the composite and the component event probabilities; compare (16.2).

Estimators for the point-wise standard errors of the empirical cumulative incidence functions are given, for example, by Beyersmann et al. (2012) and by Pintilie (2006). Both authors also provide formulas for calculating confidence intervals based on a $ln(-ln(\cdot))$ transformation to guarantee that the confidence interval lays within [0; 1]. The empirical cumulative incidence function and its point-wise confidence intervals can be computed by means of the `R` package `etm` by Allignol et al. (2011). An `R` code based on this package providing a direct comparison of the empirical cumulative incidence functions at a fixed time point to the estimated all-cause and cause-specific hazard ratios for a given data set is provided in Chap. 18. This `R` code also shows plots of the empirical cumulative incidence functions over time.

16.1.2 Estimation by the Kaplan-Meier Approach

In the previous section, the Kaplan-Meier estimator $\widehat{S}_{CE}(t)$ was introduced as a nonparametric estimator for the composite event time distribution $S_{CE}(t)$. The recommended estimator for the composite event probability was consequently given as $\widehat{P}_{CE}(t) := 1 - \widehat{S}_{CE}(t)$. It might seem intuitive at the first glance to apply the standard Kaplan-Meier concept to define point estimators and standard errors for the component-specific event probabilities as well. In fact, this approach is very often used in clinical applications; perhaps it is even the most common way of reporting event probabilities. However, the Kaplan-Meier approach relates to an estimator for the event probability only in case that no competing events are present; compare, for example, Rauch et al. (2014) and Pintilie (2006). As discussed before, the components of a composite endpoint, however, do define competing causes of failure. Whenever the components are modeled separately, the underlying multistate model is based on k competing causes of failure, and therefore the requirements for the Kaplan-Meier approach are no longer given. As long as only the composite event probability is considered, all competing causes of failure are summarized within one event class of interest. Therefore, the multistate model with one initial stage and k competing causes of failure simplifies to a two-stage model with one initial state and one final state. This implies that no competing risk scenario occurs as long as only the composite endpoint is considered. Intuitively, this no longer holds true if other relevant competing event types are present which are not included in the composite.

In the following, the point estimators for the component event probabilities based on the common Kaplan-Meier will be introduced in order to demonstrate the bad performance of these estimators. The naive way to apply the Kaplan-Meier approach to a specific cause of failure $l,\ l = 1, \ldots, k$, is given by

$$\widetilde{P}^{KM}_{EP_l}(t) := 1 - \prod_{t_{(i)} \leq t} \left(1 - \frac{\Delta N_{EP_l}(t_{(i)})}{Y(t_{(i)})}\right),\ t \geq 0,\ i = 1, \ldots, n, \tag{16.8}$$

where the index KM stands for "Kaplan-Meier" and again $t_{(1)} \leq t_{(2)} \leq \cdots \leq t_{(n)}$ denotes the ordered observed event or censoring times of the individual patients, $\Delta N_{EP_l}(t_{(i)})$ refers to the total number of observed events of type l at time $t_{(i)}$, and $Y(t_{(i)})$ denotes the set of patients under risk just prior to time $t_{(i)}$. To compare the estimator $\widetilde{P}^{KM}_{EP_l}(t)$ given in (16.8) to the empirical cumulative incidence function $\widehat{P}_{EP_l}(t)$ given in (16.7) in the previous section, we will reformulate $\widetilde{P}^{KM}_{EP_l}(t)$. For a given time point $t,\ t \geq 0$, let $m := \max\left\{i = 1, \ldots, n : t_{(i)} \leq t\right\}$ be the index of the largest observed event or censoring time which is smaller or equal than t. Then for $l = 1, \ldots, k$ the estimator $\widetilde{P}^{KM}_{EP_l}(t)$ can be written as

$$\begin{aligned}
\widetilde{P}^{KM}_{EP_l}(t) &= \widetilde{P}^{KM}_{EP_l}(t_{(m)}) \\
&= 1 - \prod_{t_{(i)} \leq t_{(m)}} \left(1 - \frac{\Delta N_{EP_l}(t_{(i)})}{Y(t_{(i)})}\right) \\
&= 1 - \left(1 - \frac{\Delta N_{EP_l}(t_{(m)})}{Y(t_{(m)})}\right) \prod_{t_{(i)} \leq t_{(m-1)}} \left(1 - \frac{\Delta N_{EP_l}(t_{(m-1)})}{Y(t_{(m-1)})}\right)
\end{aligned}$$

$$= 1 - \left(1 - \frac{\Delta N_{EP_l}(t_{(m)})}{Y(t_{(m)})}\right) \cdot \left(1 - \widetilde{P}^{KM}_{EP_l}(t_{(m-1)})\right)$$

$$= \widetilde{P}^{KM}_{EP_l}(t_{(m-1)}) + \frac{\Delta N_{EP_l}(t_{(m)})}{Y(t_{(m)})} \cdot \left(1 - \widetilde{P}^{KM}_{EP_l}(t_{(m-1)})\right)$$

$$= \sum_{t_{(i)} \leq t} \left(1 - \widetilde{P}^{KM}_{EP_l}(t_{(i-1)})\right) \cdot \frac{\Delta N_{EP_l}(t_{(i)})}{Y(t_{(i)})}$$

$$\overset{(*)}{\geq} \sum_{t_{(i)} \leq t} \widehat{S}_{CE}(t_{(i-1)}) \cdot \frac{\Delta N_{EP_l}(t_{(i)})}{Y(t_{(i)})}$$

$$= \widehat{P}_{EP_l}(t)$$

(compare Pintilie (2006)), where the inequation $(*)$ is easily verified by comparing (16.5) and (16.8). Note that $\widetilde{P}^{KM}_{EP_l}(t) = \widehat{P}_{EP_l}(t)$ only holds true if all events observed within the study correspond to the event type l under investigation. As a consequence, $\widetilde{P}^{KM}_{EP_l}(t)$ systematically overestimates the true event probability. In particular, the component event rate estimators $\widetilde{P}^{KM}_{EP_l}(t), l = 1, \ldots, k$, are no longer directly connected to the corresponding estimator $\widehat{P}_{CE}(t)$ of the composite endpoint, that means

$$\widehat{P}_{CE}(t) \leq \sum_{l=1}^{k} \widetilde{P}^{KM}_{EP_l}(t),$$

which again illustrates that the Kaplan-Meier-based estimates $\widetilde{P}^{KM}_{EP_l}(t),\ l = 1, \ldots, k$, have no meaningful interpretation. In conclusion, we recommend that the Kaplan-Meier approach should not be used to estimate the component-specific event probabilities.

In clinical applications, the Kaplan-Meier approach is often not restricted to the observed *first* events of type $l,\ l = 1, \ldots, k$, but is applied to all patients with a type-l event, irrespective of the fact whether events of other types occurred before. We strongly discourage to use this approach. Apart from the general criticism of the Kaplan-Meier approach in the presence of competing risks, this method additionally ignores that the risk to experience a subsequent event usually depends on the time point and the number of previous events.

16.1.3 Estimation by Relative Event Frequencies

The most naïve way to estimate the event probability related to a specific event type is by means of the corresponding relative event frequency obtained by dividing the number of first events of type l which occur within the study's observational period f by the total number of patients which is given by

$$\widetilde{P}^{RFF}_{EP_l}(f) := \frac{N_{EP_l}(f)}{n}, \; l = 1, \ldots, k, \tag{16.9}$$

where the index RFF stands for "relative frequency of first events" and $N_{EP_l}(f)$ refers to the total number of observed first events of type l up to time f. This approach ignores differences in the individual follow-up times. When relating the absolute number of patients who experience an event to the total number of patients, it is implicitly assumed that all patients are at risk for the event of interest at every time point within the observational period. However, whenever a patient leaves the study due to death or censoring, the risk set is reduced. As a consequence, the relative event frequency cannot be considered as an appropriate estimator for the event probability. Although this is basic knowledge from survival theory, this misuse of the relative event frequencies is often found in medical publications. It should therefore be explicitly stated that reporting of the relative event frequencies is not recommended.

As before, in clinical applications the above approach is often not restricted to the observed *first* events of type l, $l = 1, \ldots, k$, but refers to the number of patients with a type-l event irrespective of the fact whether events of other types occurred before. In this case, the relative event frequency is given as

$$\widetilde{P}^{RF}_{EP_l}(f) := \frac{M_{EP_l}(f)}{n}, \tag{16.10}$$

where the index RF stands for "relative frequency" and $M_{EP_l}(f)$ refers to the total number of observed events of type l up to time f. Note that the same criticism which applies to $\widetilde{P}^{RFF}_{EP_l}(f)$ also holds true for $\widetilde{P}^{RF}_{EP_l}(f)$. In addition, the event-specific relative frequencies $\widetilde{P}^{RF}_{EP_l}(f)$ do not sum up to the composite relative frequency to observe an event of *any* type which would hold true when considering only the first events, that is

$$\widetilde{P}^{RFF}_{CE}(f) := \frac{N_{CE}(f)}{n} = \sum_{l=1}^{k} \frac{N_{EP_l}(f)}{n} = \sum_{l=1}^{k} \widetilde{P}^{RFF}_{EP_l}(f),$$

whereas

$$\widetilde{P}^{RF}_{CE}(f) := \frac{M_{CE}(f)}{n} \leq \sum_{l=1}^{k} \frac{M_{EP_l}(f)}{n} = \sum_{l=1}^{k} \widetilde{P}^{RF}_{EP_l}(f),$$

with $N_{CE}(f)$ referring to the total number of observed first events of any type up to time f and $M_{CE}(f)$ denoting the total number of observed events of any type up to time f. This illustrates that there is no meaningful interpretation of the estimators

$\widetilde{P}^{RF}_{EP_l}(f)$, $l = 1, \ldots, k$. In addition, this method ignores that the risk to experience a subsequent event usually depends on the time point and the number of previous events.

In conclusion, relative event frequencies should generally not be reported for a descriptive analysis of a time-to-event endpoint.

16.2 Reporting of Estimated Hazard Ratios

Although comparing event probabilities between groups might seem to be an intuitive way for a descriptive evaluation of the components, it has been outlined above that in the presence of competing risks, reporting of estimated event probabilities must be done with care for the following reasons:

1. Cause-specific event probabilities cannot be interpreted separately without simultaneously looking at all competing event probabilities.
2. The event probability is a time-dependent summary measure.
3. Well-established estimators of cause-specific event probabilities are strongly biased.

In contrast to 1., the aim of a descriptive analysis of the components is to assess how the treatment under investigation influences each of the single components. For this task component-specific effect measures are required which can be interpreted separately. This is the main motivation to base a descriptive analysis of the components on hazard ratios rather than on event probabilities. In addition, as long as the proportional hazards assumption holds true (which is, however, a critical assumption), the hazard ratio is a time-independent effect measure.

As introduced in Chap. 5 in Part II of the book, the all-cause hazard ratio referring to the composite endpoint is given as

$$\theta_{CE}(t) := \frac{\lambda^C_{CE}(t)}{\lambda^I_{CE}(t)}, \; t \geq 0, \tag{16.11}$$

and the cause-specific hazard ratios referring to the components are given by

$$\theta_{EP_l}(t) := \frac{\lambda^C_{EP_l}(t)}{\lambda^I_{EP_l}(t)}, \; t \geq 0, \; l = 1, \ldots, k. \tag{16.12}$$

In there, it was also stated that the proportional hazards assumption can only hold true simultaneously for the composite and all components if the underlying baseline hazards are equal. As this is a rather strong assumption, the hazard ratios $\theta_{CE}(t)$ and $\theta_{EP_l}(t)$, $l = 1, \ldots, k$, are not necessarily constant in time. It is therefore strongly recommended to additionally report the study's observational period when providing hazard ratios as descriptive summary measures. In the following sections, the task will again be to define adequate point estimators and standard errors for $\theta_{CE}(t)$ and $\theta_{EP_l}(t)$, $l = 1, \ldots, k$. As before, note that estimated hazard ratios should generally be reported along with standard errors or confidence intervals to assess the precision of the estimates.

16.2.1 Estimation by the Cox-Model Restricted to First Events

The most common way to estimate the all-cause hazard ratio and its standard error is by means of the well-known Cox-model (Cox 1972) which assumes that the all-cause hazard function related to the composite endpoint can be modeled as

$$\lambda_{CE,i}(t, X) = \lambda_{CE,0}(t) \cdot \exp(\beta_{CE} \cdot X_i), \ t \geq 0, \ i = 1, \ldots, n, \tag{16.13}$$

where X_i is a dichotomous covariate which equals 1 whenever the ith patient belongs to the intervention and which equals 0 whenever the ith patient belongs to the control. The Cox-model is semi-parametric as the baseline hazard $\lambda_{CE,0}(t)$ remains unspecified. The regression parameter β is estimated by a partial maximum likelihood approach. An estimator for the all-cause hazard ratio is then given as

$$\widehat{\theta}_{CE} = \frac{1}{\exp(\widehat{\beta}_{CE})}. \tag{16.14}$$

Equivalently, the cause-specific hazards for the components $EP_l, l = 1, \ldots, k$, can be modeled as

$$\lambda_{EP_l,i}(t, X) = \lambda_{EP_l,0} \cdot \exp(\beta_{EP_l} \cdot X_i), \ t \geq 0, \ i = 1, \ldots, n. \tag{16.15}$$

Estimators for the cause-specific hazard ratios can then also be obtained from a partial maximum likelihood approach

$$\widehat{\theta}_{EP_l} = \frac{1}{\exp(\widehat{\beta}_{EP_l})}, \ l = 1, \ldots, k. \tag{16.16}$$

Estimators for the corresponding standard errors are, for example, discussed by Beyersmann et al. (2012). Machin et al. (2006) show the formula for a corresponding confidence interval. As mentioned above, an R code for comparing the empirical cumulative incidence functions at a given time point to the all-cause and cause-specific hazard ratios estimated from a Cox-model restricted to first events is provided in Chap. 18.

The Cox-model assumes that the group-specific hazards are proportional and thus that the hazard ratio is constant in time. It has already been discussed in Chap. 5 in Part II of this book that the proportional hazards assumption for a composite endpoint and its components defines a rather strong assumption, as the all-cause hazard of the composite is given by the sum of the cause-specific hazards related to the single components. Whenever the baseline hazards of the components are not all equal, the proportional hazards assumption for the composite is thus violated. As a consequence, in this case the hazard ratio defines a time-dependent effect measure, and the observational period must be reported for a meaningful interpretation. Instead of reporting standard hazard ratios, the average hazard ratio of Kalbfleisch and Prentice (1981) was introduced in Chap. 5 in Part II of this book. The average hazard ratio could also be reported as an alternative effect measure. The average hazard ratio is a time-dependent effect measure as well. Depending on the underlying weighting function as introduced in Chap. 5, the average hazard ratio better reflects differences between groups occurring early in time.

When modeling the all-cause hazards related to the composite endpoints, every cause of failure corresponds to an event of interest. In contrast, when modeling cause-specific hazards for the single components, it should be kept in mind that the cause-specific hazard refers to the instantaneous risk to experience a *first* event of type $l,\ l = 1, \ldots, k$. Thus, the time from start until the occurrence of the first event is measured. If this first event does not correspond to an event of type l, then the event time is censored. This approach allows an unbiased estimation of the cause-specific hazard ratios. However, it comes along with a potentially high loss in information as all events after the first are ignored. In the following sections, alternative options which allow to incorporate the information from subsequent events are discussed.

16.2.2 Estimation by the Cox-Model Not Restricted to First Events

In clinical applications, the Cox-model is often not restricted to assess the time until the occurrence of a *first* event of type $l,\ l = 1, \ldots, k$, but refers to the time a type-l event occurs irrespective of the fact whether events of other types occurred before. This approach allows to incorporate all observed events of interest, even those with a previous event. However, the hazard ratio which is modeled by this approach does not necessarily coincide with the cause-specific hazard ratio. This is due to the fact that the instantaneous risk to experience an event of a special type usually changes when another disease-related event occurs before. Even worse, the magnitude of this change can depend on the time point the previous event occurs. As a consequence, the event time distribution to assess the time until the occurrence of a type-l event is usually not the same as the event time distribution to experience a first event of type $l,\ l = 1, \ldots, k$. When incorporating events occurring after the first, the resulting treatment effect corresponds to a "total hazard ratio" which is a mixture of the treatment's so-called "direct" and "indirect" effects; compare also Rauch et al. (2016) and Cheung et al. (2010). While the direct effect expresses the direct influence of the investigated treatment on the instantaneous risk for an event of type l, the indirect effect captures an additional risk reduction which is caused by preventing or prolonging previous events of other types that would have an adverse effect on the risk for a subsequent type-l event. In conclusion, a Cox-model that ignores disease-related previous events models another treatment effect than a Cox-model that is restricted to first events which models the cause-specific hazard ratio. The total hazard ratio resulting from the first mentioned approach also delivers a meaningful treatment effect measure but has to be interpreted differently. While the cause-specific hazard ratio assesses only the treatment's direct effect, the total hazard ratio assesses a mixture of the direct and indirect effects. As the total hazard ratio is influenced by several effects, its value might be more difficult to interpret. However, a clear advantage is that all observed events of interest are taken into account and no information is lost. As a consequence, the precision of the treatment effect estimator is higher. In the special case that the instantaneous risk to experience an event of a special type is not influenced by previous disease-related events of other type, there is no indirect effect. In this special but unlikely case, both approaches model the same treatment effect given as the cause-specific hazard ratio. However, the precision of the corresponding estimators remains different as the number of underlying events to be taken into account is different within the two approaches.

16.2.3 Estimation by the Cox-Model with Time-Dependent Covariates

So far, we have introduced two different hazard ratio concepts to express the treatment effect related to a single component. While the cause-specific hazard ratios have an intuitive interpretation, an unbiased estimation of the cause-specific hazard ratios requires restriction to the set of first events only which can imply a considerable loss of information. An alternative is to ignore disease-related events of other types which occur before the event of interest. The resulting treatment effect then no longer refers to the cause-specific hazard ratio but to a total hazard ratio combining direct and indirect effects. Although this also defines a valid strategy, the total hazard ratio might be more difficult to interpret. Hence, the question is whether it is possible to include the information from subsequent events of interest while still quantifying the treatment effect by the more intuitive cause-specific hazard ratio. A possible way to do so is to extend the standard Cox-model by time-dependent covariate(s) which express the occurrence of other disease-related events. Thus, for $l = 1, \ldots, k$, the cause-specific hazards for the components are modeled as

$$\lambda_{EP_l,i}(t, X) = \lambda_{EP_l,0} \cdot \exp(\beta_{EP_l} \cdot X_i + \beta_{ODR} \cdot Z_i(t)),\ t \geq 0,\ i = 1, \ldots, n, \quad (16.17)$$

where $Z_i(t)$ is a time-dependent dichotomous covariate expressing the occurrence of another disease-related event which is not the event type of interest and β_{ODR} is the corresponding regression parameter, where the index ODR stands for "other disease-related events." If a disease-related event occurs at time t before the event of interest, the covariate $Z_i(s)$ equals 0 for all $s < t$ and 1 for all $s \geq t$. Note that if the component EP_l is currently under investigation, then all other components of type other than l define the disease-related events which are captured by the covariate $Z_i(t)$. Of course, it is also possible to incorporate several time-dependent covariates to model the influence of the different components separately. However, this is only reasonable if the underlying sample size is adequately large.

The model given in (16.17) assumes that disease-related events that occur before the event of interest have an influence on the underlying cause-specific *hazard* but not on the cause-specific *hazard ratio*. In real data situations, however, the underlying hazard ratio might also be influenced by the occurrence of previous events. It is important to note that in the latter case, the above approach no longer models the cause-specific hazard ratio but again a mixture of direct and indirect effects. Note that defining a model which remains intuitive to interpret and easy to apply but at the same time captures the influence of covariates on the hazard ratio is difficult. Therefore, although knowing about the model limitations, the extended Cox-model incorporating time-dependent covariates is an intuitive and attractive option to include the information from all events of interest without ignoring the influence of other disease-related events.

16.3 Incorporating Multiple Events per Patient

16.3.1 Reporting of Absolute Event Frequencies

In many clinical trials, the components of a composite time-to-event endpoint are simply evaluated by providing the absolute numbers of observed events for each event type under consideration within the different treatment groups. By this, it is also possible to report several events per patient and thus events occurring after the first are fully counted. However, deducing conclusions about the component-specific treatment effects by assessing the absolute number of events per treatment arm is difficult or even impossible for the following reasons.

The absolute event frequencies depend on the observation period. Thus, the time point of assessment must be reported for a meaningful interpretation. However, the patients' individual follow-up periods usually differ due to the underlying recruitment scheme or due to different censoring mechanisms. As a consequence, the number of patients under risk does not remain the same over time. Therefore, the expected number of events per patient naturally differ even if all patients within the same treatment group are exposed to the same instantaneous risk to experience the event of interest. Therefore, the absolute number of observed events is difficult to interpret as a low number of events may be due to an effective treatment but also to a shorter follow-up. A comparison between the absolute numbers of events between different treatment arms is only meaningful if the sample sizes per treatment arm are equal and if the mean follow-up is similar in each treatment group. This usually holds true as long as the event type under investigation is only subject to non-informative censoring. However, when investigating a single component, all other components define competing risks. Therefore, the censoring process becomes treatment-related. For example, if more people die in the control arm than in the intervention arm, the mean follow-up in the control group will be shorter. As a consequence, fewer nonfatal events like myocardial infarction might be observed in the control arm, and thus the absolute number of myocardial infarctions will be lower. However, this is not due to a protective effect in the control group but to a shorter follow-up caused by a harmful competing event.

Another shortcoming of reporting absolute event frequencies is that events belonging to the same patient and events belonging to different patients are mixed up. In general, multiple events that occur within a single patient are correlated, whereas events between different patients are naturally unrelated. Thus, there exists a cluster structure of the data which is ignored when reporting the absolute number of events irrespective of the fact to which patient the events belong. For example, if a single patient is seriously ill, a high number of adverse events might occur for this individual patient, whereas only a few events occur for all other patients within the same treatment group. The absolute event frequency for this treatment group is then relatively high although this is caused only by a single patient. In addition, the competing risk situation described above will also influence the absolute number of observed events. If a high number of patients die early in the control group, then the number of nonfatal events will be naturally low in this group. Therefore, a lower absolute event frequency does not necessarily imply a beneficial effect.

In conclusion, the absolute number of observed events should not be considered to investigate component-specific treatment effects. However, investigating the absolute number of events in comparison to the number of observed first events can give an impression about the amount of information which is lost by analyzing the time-to-first-event only. Therefore, as an additional information, the absolute event frequencies might be reported. In this case, the observation time point as well as the median or mean observation times per group should be provided in addition.

16.3.2 Reporting of Mean Number of Events per Patient Year

To overcome the problem of individual follow-up periods per patient, a possible approach is to provide the mean number of events per patient year for each treatment arm. To do so, the total number of patient years per group is determined by summing up all individual follow-up times for all patients. Subsequently, in each group the absolute event frequency is divided by the number of patient years. The ICH E9 Guideline (ICH 1998) comments on the problem of individual follow-up periods as follows

> The incidence of a certain adverse event is usually expressed in the form of a proportion relating number of subjects experiencing events to number of subjects at risk. However, it is not always self-evident how to assess incidence. For example, depending on the situation the number of exposed subjects or the extent of exposure (in person-years) could be considered for the denominator.

While the mean number of events per patient year overcomes the problem of differences in the individual follow-up times, the problems related to the cluster structure of the data and to the competing risk scenario remain the same as for the absolute event frequencies. Therefore, a lower mean number of events per patient year does not necessarily imply a better outcome. Despite these shortcomings, the mean number of events per patient year defines an intuitive way to report multiple events per patient and should clearly be preferred over reporting simple absolute event frequencies. If the mean number of events per patient year is reported, we recommend to additionally investigate other treatment effect measures which focus on a single event per patient in order to avoid serious misinterpretations due to competing causes of failure or due to an underlying non-identified cluster structure of the events.

In Chap. 18 an `R` code for calculating the mean number of events per patient year in each treatment group for a given data set is provided.

References

Allignol, A., Schumacher, M., & Beyersmann, J. (2011). Empirical transition matrix of multi-state models: The etm package. *Journal of Statistical Software, 38*, 1–15.

Beyersmann, J., Allignol, A., & Schumacher, M. (2012). *Competing risks and multistate models with R*. New York: Springer.

Cheung, Y. B., Xu, Y., Tan, S. H., Cutts, F., & Milligan, P. (2010). Estimation of intervention effects using first or multiple episodes in clinical trials: The Andersen-Gill model re-examined. *Statistics in Medicine, 29*, 328–336.

Cox, D. (1972). Regression models and life tables. *Journal of the Royal Statistical Society B, 34*, 187–220.

ICH. (1998). *Statistical principles for clinical trials - E9*. ICH. http://www.ich.org/fileadmin/Public_Web_Site/ICH_Products/Guidelines/Efficacy/E9/Step4/E9_Guideline.pdf. Accessed April 9, 2018.

Kalbfleisch, J. D., & Prentice, R. L. (1981). Estimation of the average hazard ratio. *Biometrika, 68*, 105–112.

Kaplan, E. L., & Meier, P. (1958). Nonparametric estimation from incomplete observations. *Journal of the American Statistical Association, 53*, 457–481.

Machin, D., Cheung, Y. B., & Parmer, M. K. B. (2006). *Survival analysis: A practical approach* (2nd ed.). New York: Wiley.

Pintilie, M. (2006). *Competing risks: A practical perspective*. Chichester: John Wiley & Sons.

Rauch, G., Kieser, M., Binder, H., Bayes-Genis, A., & Jahn, A. (2016). Time-to-first-event versus recurrent-event analysis - Points to consider for selecting a meaningful analysis strategy in clinical trials with composite endpoints. DOI: 10.1007/s00392-018-1205-7.

Rauch, G., Kieser, M., Ulrich, S., Doherty, P., Rauch, B., Schneider, S., et al. (2014). Competing time-to-event endpoints in cardiology trials: A simulation study to illustrate the importance of an adequate statistical analysis. *European Journal of Preventive Cardiology, 21*, 74–80.

Chapter 17
Supplementary Confirmatory Analyses of the Components

As discussed above, clinical trials using a composite endpoint as primary outcome are often difficult to interpret as the single components may show varying or even opposite treatment effects. Although a descriptive analysis of the single components is the current standard and can reveal those problems, this approach does not provide any confirmatory evidence and thus gives no guidance for a definite conclusion on the treatment efficacy. Methods for a confirmatory analysis of a composite endpoint and one main component have been presented in Part III of this book. In there, the focus was to define an efficacy claim based on a multiple test problem and to calculate the sample size such that there is sufficient power to reach this claim. However, a confirmatory analysis of the components can still be of interest, even if the underlying multiple test problem does not correspond to the formal efficacy claim for which the trial is powered. In this chapter, the focus lies on simple confirmatory multiple test strategies for the components which provide a reasonable chance to get additional confirmatory evidence without the need to increase sample size.

17.1 Gatekeeping Strategies

Additional confirmatory claims for the components can be implemented by using the primary efficacy endpoint as a gatekeeper. The components are thus tested if and only if the main null hypothesis has been rejected. Although the power to reject further null hypotheses related to the components in addition to the null hypothesis related to the composite is naturally lower than the anticipated power for the composite null hypothesis alone, the probability to gain further confirmatory evidence on at least some components can be considerable (Schüler et al. 2014). Furthermore, using the composite as a gatekeeper, the power for testing the composite null hypothesis is not affected by the additional option to test further component null hypotheses. Therefore, this strategy can generally be recommended. Note that the gatekeeping test problem must not necessarily rely on a single endpoint. It is also possible to define a multiple test problem as a

G. Rauch et al., *Planning and Analyzing Clinical Trials with Composite Endpoints*, Springer Series in Pharmaceutical Statistics,
https://doi.org/10.1007/978-3-319-73770-6_17

gatekeeper, for example, an intersection-union test assessing the composite and the most relevant component. The subsequent multiple test strategy for the remaining components can be chosen arbitrarily. In the literature, a variety of sequentially rejective approaches are discussed which are based on weighted Bonferroni-type tests. The most simple one is the standard Bonferroni-Holm method (Holm 1979); extensions are given by advanced gatekeeping procedures (Westfall and Krishen 2001; Bauer et al. 1998; Dmitrienko et al. 2003; Chen et al. 2005), fixed sequence tests (Wiens 2003), and fallback procedures (Wiens and Dmitrienko 2005). The procedures address the endpoints of interest with respect to their clinical importance and their relationship between each other. Bretz et al. (2009) proposed a graphical method to illustrate different sequentially rejective test procedures which allow to reallocate the local significance levels of hypotheses which have already been rejected. For illustrative purposes, we present in the following two simple multiple test strategies for an additional confirmatory assessment of the components.

17.1.1 Composite Endpoint as Gatekeeper Followed by Bonferroni-Holm Procedure

In cases where the composite endpoint consists of components that are either similar in their expected effect sizes or the effect sizes are difficult to predict, a suitable multiple test procedure would be a combination of gatekeeping and the Bonferroni-Holm procedure (Westfall and Krishen 2001; Holm 1979), where the composite endpoint defines a gatekeeper for the component tests. If the composite null hypothesis is rejected at the full level α, the components are tested in the order of their observed p-values starting with the smallest. The smallest p-value is compared to the Bonferroni-adjusted level of α/k, where k defines the number of components. If the related null hypothesis can be rejected, the second smallest p-value is compared to a level of $\alpha/(k-1)$ and so on. If the composite null hypothesis is not rejected, all component null hypotheses cannot be rejected as well. By this strategy, it is likely to gain additional confirmatory evidence for the components related to the largest effects without the need to prespecify which components are expected to yield these effects and without compromising the power for the composite.

17.1.2 Hierarchical Testing of the Composite and the Components

If a composite endpoint is used for which the effects of the components are expected to be of considerably different magnitude, an alternative multiple test strategy would be hierarchical testing. The hierarchy of test problems starts with testing the composite endpoint which again defines a gatekeeper, followed by level-α tests for the components in order of their expected effect sizes starting with the largest. The procedure stops as soon as a null hypothesis cannot be rejected at level α, and all null hypotheses that are lower in the hierarchy are accepted as well. The advantage of this predefined ordering is that all tests can be performed at the full significance level α which provides a power advantage compared to the Bonferroni-Holm approach as long as the assumed ordering of the effect sizes is correct. However, it should be noted that the gain in information of this multiple test strategy strongly depends on the correct ordering of hypotheses. If the information on the expected effect sizes is limited in the planning stage, the Bonferroni-Holm approach presented above should be preferred.

References

Bauer, P., Röhmel, J., Maurer, W., & Hothorn, L. (1998). Test strategies in multi-dose experiments including active control. *Statistics in Medicine, 17*, 2133–2146.

Bretz, F., Maurer, W., Brannath, W., & Posch, M. (2009). A graphical approach to sequentially rejective multiple test procedures. *Statistics in Medicine, 28*, 586–604.

Chen, X., Luo, X., & Caprizzi, T. (2005). The application of enhanced gatekeeping strategies. *Statistics in Medicine, 24*, 1385–1397.

Dmitrienko, A., Offen, W. W., & Westfall, P. H. (2003). Gatekeeping strategies for clinical trials that do not require all primary effects to be significant. *Statistics in Medicine, 22*, 2387–2400.

Holm, S. (1979). A simple sequentially rejective multiple test procedure. *Scandinavian Journal of Statistics, 6*, 65–70.

Schüler, S., Mucha, A., Doherty, P., Kieser, M., & Rauch, G. (2014). Easily applicable multiple test procedures to improve the interpretation of clinical trials with composite endpoints. *International Journal of Cardiology, 175*, 126–132.

Westfall, P. H., & Krishen A. (2001). Optimally weighted, fixed sequence, and gatekeeping multiple test procedures. *Journal of Statistical Planning and Inference, 99*, 25–40.

Wiens, B. L. (2003). A fixed sequence Bonferroni procedure for test multiple endpoints. *Pharmaceutical Statistics, 2*, 211–215.

Wiens, B. L., & Dmitrienko, A. (2005). The fallback procedure for evaluating a single family of hypotheses. *Journal of Biopharmaceutical Statistics, 15*, 929–942.

Chapter 18
Related Software Code

This chapter provides exemplary source code written in the software R implementing different approaches outlined in Part V of this book.

18.1 Calculating Cause-Specific Hazard Ratios and Cumulative Incidence Functions

```
library(etm)
library(survival)

###################################################
# Main function to plot and calculate the         #
# empirical cumulative incidence functions per    #
# group for a given time point along with         #
# estimators for all-cause and cause-specific     #
# hazard ratios all with the corresponding        #
# 95% confidence intervals for a given data set   #
###################################################

###################################################
# Parameters to specify                           #
###################################################
#
#
# Before you start the program, it is necessary to
# import your data into R.
#
# Required structure of the data set "data":
#
# 1.column (id)      = Patient ID
# 2.column (entry)   = Entry time; if set to 0
#                      variables "times" correspond
#                      to event time
# 3.column (times)   = Out time; if "entry time"
#                      set to 0 variable "times"
#                      correspond to event time
```


G. Rauch et al., *Planning and Analyzing Clinical Trials with Composite Endpoints*, Springer Series in Pharmaceutical Statistics,
https://doi.org/10.1007/978-3-319-73770-6_18

```
# 4.column (status)   = Event indicator indicating
#                       type of event
#                       (1,2,3; 0 for censoring)
# 5.column (statusCE)= Binary event indicator for
#                       CE (Composite)
# 6.column (group)    = Group allocation
#---------------------------------------------------
# Other required parameters
#---------------------------------------------------
# t                   = Timepoint CIF is evaluated
####################################################

CIF_HR<-function(data,t){


  #----------------------------------------------#
  # Calculate all-cause hazard ratio and 95% CI #
  # for the composite endpoint                   #
  #----------------------------------------------#


  cox_ce<-coxph(Surv(entry,times,status!=0)~group,
               data)
  HR_ce<-1/exp(cox_ce$coefficients)
  HR_ce_l<- exp(-cox_ce$coefficients-
      qnorm(0.975)*summary(cox_ce)$coefficients[3])
  HR_ce_u<- exp(-cox_ce$coefficients+
      qnorm(0.975)*summary(cox_ce)$coefficients[3])



  #----------------------------------------------#
  # Calculate point estimators and 95% CIs for  #
  # the CIFs of the composite endpoint           #
  #----------------------------------------------#


  CIF_ce<-etmCIF(Surv(times,status!=0)~group,
                data,
                etype=status_ce,
                failcode=1)

  #----------------------------------------------#
  # Plot the CIFs of the composite endpoint      #
  #----------------------------------------------#


  plot.new()
  plot(CIF_ce,
       col = c(4,4),
       lty = c(1,2),
       xlim = c(0, t),
       ylim = c(0,1),
       curvlab = c("CIF CE, Control",
```

```
                "CIF CE, Intervention"))
par(new=FALSE)

#-------------------------------------------#
# Intervention group                        #
#-------------------------------------------#

times<-CIF_ce$'group=1'$time
timepoint<-max(times[times<=t])
CIF_ce_g1<-CIF_ce$'group=1'$est[, ,
                        which(times==timepoint)]

arg_g1_ce<-qnorm(0.975)*sqrt(CIF_ce$'group=1'$cov[
  "0 1","0 1",which(times==timepoint)])/(
    (1-CIF_ce_g1[1,2])*(log(1-CIF_ce_g1[1,2])))

CIF_ce_g1_u<-1-(1-CIF_ce_g1[1,2])^(exp(-arg_g1_ce))
CIF_ce_g1_l<-1-(1-CIF_ce_g1[1,2])^(exp(+arg_g1_ce))

#-------------------------------------------#
# Control group                             #
#-------------------------------------------#

times<-CIF_ce$'group=0'$time
timepoint<-max(times[times<=t])
CIF_ce_g0<-CIF_ce$'group=0'$est[, ,
                        which(times==timepoint)]

arg_g0_ce<-qnorm(0.975)*sqrt(CIF_ce$'group=0'$cov[
  "0 1","0 1",which(times==timepoint)])/(
    (1-CIF_ce_g0[1,2])*(log(1-CIF_ce_g0[1,2])))

CIF_ce_g0_u<-1-(1-CIF_ce_g0[1,2])^(exp(-arg_g0_ce))
CIF_ce_g0_l<-1-(1-CIF_ce_g0[1,2])^(exp(+arg_g0_ce))


#-------------------------------------------#
# Calculate cause-specific hazard ratios    #
# and 95% CI for the composite endpoint     #
#-------------------------------------------#


cox_ep1<-coxph(Surv(times,status==1)~group,data)
HR_ep1<-1/exp(cox_ep1$coefficients)
HR_ep1_l<- exp(-cox_ep1$coefficients-
     qnorm(0.975)*summary(cox_ep1)$coefficients[3])
HR_ep1_u<- exp(-cox_ep1$coefficients+
     qnorm(0.975)*summary(cox_ep1)$coefficients[3])

cox_ep2<-coxph(Surv(times,status==2)~group,data)
HR_ep2<-1/exp(cox_ep2$coefficients)
HR_ep2_l<- exp(-cox_ep2$coefficients-
     qnorm(0.975)*summary(cox_ep2)$coefficients[3])
HR_ep2_u<- exp(-cox_ep2$coefficients+
```

```
     qnorm(0.975)*summary(cox_ep2)$coefficients[3])

cox_ep3<-coxph(Surv(times,status==3)~group,data)
HR_ep3<-1/exp(cox_ep3$coefficients)
HR_ep3_l<- exp(-cox_ep3$coefficients-
     qnorm(0.975)*summary(cox_ep3)$coefficients[3])
HR_ep3_u<- exp(-cox_ep3$coefficients+
     qnorm(0.975)*summary(cox_ep3)$coefficients[3])



#----------------------------------------------#
# Calculate point estimators an 95% CIs for   #
# the CIFs of the components                   #
#----------------------------------------------#


CIF<-etmCIF(Surv(entry,times,status!= 0)~group,
            data,
            etype=status, failcode=1)

plot(CIF, which.cif = c(1,2,3),
     col = c(1,2,3,1,2,3),
     lty = c(1,1,1,2,2,2),
     xlim = c(0,t),
     ylim=c(0,1),
     curvlab = c("CIF EP1, Control",
                 "CIF EP2, Control",
                 "CIF EP3, Control",
                 "CIF EP1, Intervention",
                 "CIF EP2, Intervention",
                 "CIF EP3, Intervention"))

#----------------------------------------------#
# Intervention group                           #
#----------------------------------------------#

times<-CIF$`group=1`$time
timepoint<-max(times[times<=t])
CIF_g1<-CIF$`group=1`$est[, ,
                        which(times==timepoint)]

arg_g1_ep1<-qnorm(0.975)*sqrt(CIF$`group=1`$cov[
  "0 1","0 1",which(times==timepoint)])/(
    (1-CIF_g1[1,2])*(log(1-CIF_g1[1,2])))

CIF_ep1_g1_u<-1-(1-CIF_g1[1,2])^(exp(-arg_g1_ep1))
CIF_ep1_g1_l<-1-(1-CIF_g1[1,2])^(exp(+arg_g1_ep1))


arg_g1_ep2<-qnorm(0.975)*sqrt(CIF$`group=1`$cov[
  "0 2","0 2",which(times==timepoint)])/(
    (1-CIF_g1[1,3])*(log(1-CIF_g1[1,3])))
```

```
CIF_ep2_g1_u<-1-(1-CIF_g1[1,3])^(exp(-arg_g1_ep2))
CIF_ep2_g1_l<-1-(1-CIF_g1[1,3])^(exp(+arg_g1_ep2))


arg_g1_ep3<-qnorm(0.975)*sqrt(CIF$'group=1'$cov[
  "0 3","0 3",which(times==timepoint)])/(
    (1-CIF_g1[1,4])*(log(1-CIF_g1[1,4])))

CIF_ep3_g1_u<-1-(1-CIF_g1[1,4])^(exp(-arg_g1_ep3))
CIF_ep3_g1_l<-1-(1-CIF_g1[1,4])^(exp(+arg_g1_ep3))


#-------------------------------------------#
# Control group                             #
#-------------------------------------------#

times<-CIF$'group=0'$time
timepoint<-max(times[times<=t])
CIF_g0<-CIF$'group=0'$est[, ,
                         which(times==timepoint)]

arg_g0_ep1<-qnorm(0.975)*sqrt(CIF$'group=0'$cov[
  "0 1","0 1",which(times==timepoint)])/(
    (1-CIF_g0[1,2])*(log(1-CIF_g0[1,2])))

CIF_ep1_g0_u<-1-(1-CIF_g0[1,2])^(exp(-arg_g0_ep1))
CIF_ep1_g0_l<-1-(1-CIF_g0[1,2])^(exp(+arg_g0_ep1))


arg_g0_ep2<-qnorm(0.975)*sqrt(CIF$'group=0'$cov[
  "0 2","0 2",which(times==timepoint)])/(
    (1-CIF_g0[1,3])*(log(1-CIF_g0[1,3])))

CIF_ep2_g0_u<-1-(1-CIF_g0[1,3])^(exp(-arg_g0_ep2))
CIF_ep2_g0_l<-1-(1-CIF_g0[1,3])^(exp(+arg_g0_ep2))


arg_g0_ep3<-qnorm(0.975)*sqrt(CIF$'group=0'$cov[
  "0 3","0 3",which(times==timepoint)])/(
    (1-CIF_g0[1,4])*(log(1-CIF_g0[1,4])))

CIF_ep3_g0_u<-1-(1-CIF_g0[1,4])^(exp(-arg_g0_ep3))
CIF_ep3_g0_l<-1-(1-CIF_g0[1,4])^(exp(+arg_g0_ep3))


#-------------------------------------------#
# Generate output                           #
#-------------------------------------------#
```

```
  out<-matrix(c(HR_ce,HR_ep1,HR_ep2,HR_ep3,
              HR_ce_l,HR_ep1_l,HR_ep2_l,HR_ep3_l,
              HR_ce_u,HR_ep1_u,HR_ep2_u,HR_ep3_u,
              CIF_ce_g0[1,2], CIF_g0[1,2],
              CIF_g0[1,3], CIF_g0[1,4],
              CIF_ce_g0_l, CIF_ep1_g0_l,
              CIF_ep2_g0_l, CIF_ep3_g0_l,
              CIF_ce_g0_u,CIF_ep1_g0_u,
              CIF_ep2_g0_u, CIF_ep3_g0_u,
              CIF_ce_g1[1,2], CIF_g1[1,2],
              CIF_g1[1,3], CIF_g1[1,4],
              CIF_ce_g1_l, CIF_ep1_g1_l,
              CIF_ep2_g1_l, CIF_ep3_g1_l,
              CIF_ce_g1_u,CIF_ep1_g1_u,
              CIF_ep2_g1_u,CIF_ep3_g1_u),
              nrow=4,ncol=9,byrow=FALSE)
  dimnames(out) = list(c("CE","EP1","EP2", "EP3"),
                c("HR (Control/Intervention)",
                "95% CI LB",
                "95% CI UB",
                "CIF Control",
                "95% CI LB",
                "95% CI UB",
                "CIF Intervention",
                "95% CI LB",
                "95% CI UB"))

  print("CIF evaluated at time point (in years)")
  print(t)
  return(out)

}


#************************************#
# Exemplary calls                    #
#************************************#
set.seed(30)

#------------------------------------#
# Specify all-cause hazards          #
#------------------------------------#

haz_I<-0.85
haz_C<-1.85

#------------------------------------#
# Generate event indicators for two  #
# components                         #
#------------------------------------#

statusI<-rmultinom(200, 1,
```

```
                   c(0.5/haz_I,0.083/haz_I,
                     0.167/haz_I,0.1/haz_I))
statusC<-rmultinom(200, 1,
                   c(0.75/haz_C,0.833/haz_C,
                     0.167/haz_C,0.1/haz_C))


status1<-c(statusI[1,], statusC[1,])
status2<-c(statusI[2,], statusC[2,])
status3<-c(statusI[3,], statusC[3,])

#-------------------------------------#
# Composite equals 1 whenever at      #
# least one of the components is 1    #
#-------------------------------------#

entry<-rep(0,400)

status<-rep(0,400)
status_ce<-rep(0,400)
status[status1==1]<-1
status[status2==1]<-2
status[status3==1]<-3
status_ce[status!=0]<-1


data<-data.frame(id=c(1:400),
       times=c(-log(runif(200,0,1))/(haz_I),
               -log(runif(200,0,1))/(haz_C)),
       entry=entry,
       status=status,
       status_ce=status_ce,
       group=c(rep(1,200),rep(0,200)))

print("True Hazard Ratios CE, EP1,EP2,EP3")
print(c(1.85/0.85,0.75/0.5,0.833/0.083,0.167/0.167))
CIF_HR(data,3)
```

18.2 Calculating Mean Number of Events per Patient Year

```
###################################################
# Main function to calculate the mean number of   #
# events per patient-year for a given data set    #
###################################################

###################################################
# Parameters to specify                           #
###################################################
#
#
# Before you start the program, it is necessary to
# import your data into R.
```

```
#
# Required structure of the data set "data":
#
# 1.column (id)       = Patient ID
# 2.column (times)    = Individual observation time
# 3.column (nevents) = Number of observed events
#                       for individual patient
# 4.column (group)    = Group allocation
####################################################

MeanEvents_PY<-function(data){

  sum_times_g1    <-sum(data$times[data$group==1])
  sum_events_g1   <-sum(data$nevents[data$group==1])

  sum_times_g0    <-sum(data$times[data$group==0])
  sum_events_g0   <-sum(data$nevents[data$group==0])

  MEPY_g1         <-sum_events_g1/sum_times_g1
  MEPY_g0         <-sum_events_g0/sum_times_g0

  #----------------------------------------------#
  # Generate output                              #
  #----------------------------------------------#

  out<-matrix(c(MEPY_g0,MEPY_g1),
              nrow=1,ncol=2,byrow=FALSE)
  dimnames(out) = list(c(
    "Mean Number of Events per Patient-Year"),
                c("Control", "Intervention"))

  return(out)

}

#************************************#
# Exemplary calls                    #
#************************************#

set.seed(50)

#------------------------------------#
# Generate exemplary data set        #
#------------------------------------#

data<-data.frame(id=c(1:400),
       times=c(-log(runif(200,0,1))/(0.7),
               -log(runif(200,0,1))/(0.9)),
       nevents=c(rbinom(200,8,0.2),
                 rbinom(200,10,0.4)),
       group=c(rep(1,200),rep(0,200)))

print(data)
MeanEvents_PY(data)
```

Part VI
Illustrating Clinical Trial Examples

In this part of the book, the methods for planning and analyzing clinical trials with composite endpoints presented throughout this book are illustrated by means of different clinical trial examples introduced in Chap. 4 in Part I of this book. The aim of this part is to illustrate how the methods can be applied to overcome the problems related to composite endpoints and thereby maintaining their advantages. It will also be shown that there is often not only a single adequate way to plan or analyze a clinical trial with a composite endpoint. For ease of readability, the main design characteristics of the different examples stated already in Chap. 4 in Part I are repeated here. By this, the motivation for the choice of the investigated planning and analysis methods becomes more intuitive. Where appropriate, different methods are illustrated by means of the same clinical trial example in order to point out the differences and the similarities. As the original data of the exemplary trials were not available, the methods are applied to simulated data sets which try to mimic the true setting. Thereby, the focus lays on illustrating the methods for composite endpoints discussed within this book. The conclusions which can be drawn from our applications should exclusively be used to assess the performance of the different statistical methods but not to judge or to criticize the original design or the results of the clinical trial examples.

Chapter 19
Clinical Trial Examples with (Composite) Binary Endpoints

In this chapter, we discuss several original clinical trials for composite binary endpoints along with their designs, results, and possible limitations. The focus of this chapter is to propose and apply alternative planning and analysis strategies which might help to improve the performance of the design or to ease interpretation of results.

19.1 The Osteoporosis Trial

In a randomized, controlled clinical trial on a screening program for osteoporosis (Yuksel et al. 2010), the effect of a community pharmacist screening program on testing and treating of osteoporosis was assessed. Patients in the intervention group received an intensive education program on aspects of osteoporosis; patients in the control group received usual osteoporosis care in the community pharmacy. The primary endpoint was defined as a composite binary endpoint of two indicator variables. The first indicator equals 1 if a new initiation of osteoporosis medication was done; the second equals 1 whenever a bone mineral density test was performed. Note that in contrast to the other examples presented within this book, a higher event rate corresponds to a more favorable outcome in this application. The planned total sample size of 262 patients provides a power of 0.95 to detect an increase of the composite success rate from 0.2 in the usual care group to 0.4 in the intervention group for a one-sided significance level of 0.025 assuming equal group allocation.

The study showed a clear superiority of the intervention with respect to the composite (success rate of 0.22 observed in the intervention group versus 0.11 in the control group, $p < 0.0001$ when tested with the normal approximation test for rates). However, this outcome was mainly driven by the bone mineral density test endpoint (success rate 0.22 in the intervention group versus 0.10 in the control group), whereas the new medication rates were rather small (success rate of 0.05 in the intervention versus 0.02 in the placebo group).

G. Rauch et al., *Planning and Analyzing Clinical Trials with Composite Endpoints*, Springer Series in Pharmaceutical Statistics,
https://doi.org/10.1007/978-3-319-73770-6_19

19.1.1 Robust Sample Size for a Multiple Test Problem

While the original study was powered exclusively to assess the composite, additional confirmatory evidence on the single components would clearly yield a gain in information. Therefore, the composite and its components will be assessed for illustration within a hierarchical test problem, where the composite endpoint defines a gatekeeper which is tested at the full one-sided significance level of 0.025. Both components are tested at the Bonferroni-adjusted significance level 0.0125 if and only if the composite has been demonstrated to be significant. While the composite endpoint is tested for superiority, both components are tested for non-inferiority in order to exclude major adverse effects.

We begin with calculating the frequentist power for the multiple test problem using the same parameter assumptions for the composite as used in the original approach. The parameter assumptions for the components are chosen such that the anticipated effect size for the first component is higher than for the second:

$$p_{CE}^{I} = 0.4,\ p_{CE}^{C} = 0.2,$$
$$p_{NewMed}^{I} = 0.1,\ p_{NewMed}^{C} = 0.05,$$
$$p_{Test}^{I} = 0.35,\ p_{Test}^{C} = 0.175.$$

Note that the above parameter assumptions include an overlap between the components. Assuming that the anticipated event rate $p_{NewMed}^{C} = 0.05$ is correctly specified, the worst case in terms of an effect in opposite direction would be $p_{CE}^{I} - p_{CE}^{C} = -0.05$. In order to exclude this worst-case scenario in a confirmatory way, we chose $\delta_{NewMed} = 0.05$. This rather liberal margin is only acceptable as superiority for the composite is additionally required for a successful trial. Moreover, the component "new initiation of osteoporosis medication" is not a hard endpoint concerning patients' safety, so a liberal non-inferiority margin does not imply a higher risk for the patient. For the second component which refers to the performance of a bone mineral density test, the expected effects are higher, and thus a more liberal non-inferiority margin $\delta_{Test} = 0.1$ is chosen. The specific non-inferiority margins and the above parameter assumptions yield a frequentist power for rejecting all null hypotheses within the multiple test problem of 0.66 which is considerably smaller than the power for the test problem assessing exclusively the composite. The power to reject the null hypotheses for the composite and for the first component only is given by 0.66, whereas the power to reject the null hypotheses for the composite and for the second component is given by 0.95. Thus, the global power to reject all null hypotheses is exclusively reduced by the first component claim. These exemplary power calculations illustrate that the sample size which is calculated exclusively to assess the composite usually does not provide enough power to additionally assess the components. In this case, the calculation of the global power for the multiple test problem indicates that any interpretation concerning the effects for the individual component should be done with care.

However, the above power calculation for the multiple test problem does not seem to be very reliable as the parameter assumptions were not based on a strong level of data evidence. We will therefore take account of this uncertainty by replacing the fixed parameter assumptions by prior distributional assumptions:

$$\begin{aligned}
p_{CE}^{I} &\sim Tri\,[0.35; 0.45; 0.4]\,, \\
p_{CE}^{C} &\sim Tri\,[0.15; 0.25; 0.2]\,, \\
p_{NewMed}^{I} &\sim Tri\left[0.2 \cdot p_{CE}^{I}; 0.3 \cdot p_{CE}^{I}; 0.25 \cdot p_{CE}^{I}\right], \\
p_{NewMed}^{C} &\sim Tri\left[0.2 \cdot p_{CE}^{C}; 0.3 \cdot p_{CE}^{C}; 0.25 \cdot p_{CE}^{C}\right], \\
p_{Test}^{I} &\sim Tri\left[0.75 \cdot p_{CE}^{I}; p_{CE}^{I}; 0.875 \cdot p_{CE}^{I}\right], \\
p_{Test}^{C} &\sim Tri\left[0.75 \cdot p_{CE}^{C}; p_{CE}^{C}; 0.875 \cdot p_{CE}^{C}\right].
\end{aligned}$$

Note that the prior intervals for the second component are chosen such that simultaneous events in both endpoints are possible so that the overlap rates can be different from 0. The above prior distributional assumptions yield an expected power of 0.65 for rejecting all null hypotheses which is very similar to the frequentist power for the multiple test problem. Note, however, that the minimal power which corresponds to the power for the worst-case parameter constellation within the above ranges is considerably smaller with 0.21. This illustrates that looking at the frequentist power for the worst-case scenario of a certain parameter range often does not provide a realistic approach as the required sample size would be too large to be realized. The expected power takes account of the worst-case scenario, too, but it is considered as an unlikely event out of many other possible scenarios. For this reason, an expected power approach is more intuitive here. The power to reject the null hypotheses for the composite and for the first component amounts to 0.65 only, whereas the power to reject the null hypotheses for the composite and for the second component is given by 0.92. Again, the global power to reject all null hypotheses is exclusively reduced by the first component claim.

When comparing the original parameter assumptions of the trial with the observed event rates, it might even seem reasonable to define the priors such that a higher amount of uncertainty is modeled. Therefore, we additionally consider another scenario where the event rates for the composite are determined by the following priors:

$$\begin{aligned}
p_{CE}^{I} &\sim Tri\,[0.25; 0.45; 0.35]\,, \\
p_{CE}^{C} &\sim Tri\,[0.10; 0.25; 0.175]\,.
\end{aligned}$$

Taking the prior assumptions for the components as specified above, this results in an expected power of 0.6. In this case, the expected power approach yields smaller results than the frequentist approach which indicates that the parameters for the frequentist power calculation might have been chosen too optimistic. The power loss can therefore be interpreted as the price one has to pay for the higher amount of uncertainty about the parameter assumptions. For comparison, the power to reject the null hypotheses for the composite and for the first component is only 0.6, whereas the power to reject the null hypotheses for the composite and for the second component is given by 0.83.

This clinical trial example shows that initial parameter assumptions might differ considerably from the unknown true event rates which can result in a notable loss of power. Using an expected power approach to assess the composite and its components can give a more realistic impression of the sample size needed for a valid study interpretation. Depending on the level of evidence about the parameter assumptions, the expected power can be close to the frequentist power if there exists reliable inference on the parameter assumptions. However, if the uncertainty about the parameter assumptions is rather high, the expected power can be much smaller which indicates that the sample size should be increased. Moreover, this example illustrates well that the choice of an appropriate non-inferiority margin is a crucial issue of our method. When testing a hard component, e.g., death, for non-inferiority, the non-inferiority margin cannot been chosen as liberal as demonstrated in this example. A non-inferiority test problem for a hard endpoint corresponding to a small expected absolute effect size requires a conservative margin in order to guarantee the patients' safety. This will always yield to a dramatic loss in power compared to the single hypothesis test problem for the composite alone. However, this power consideration is still useful in order to judge the utility of such a trial.

This exemplary illustration of robust sample size calculation was originally published by Rauch and Kieser (2013).

19.2 The MOMS Trial

In another randomized, controlled study in pregnant women, prenatal repair of myelomeningocele was compared to repair after delivery which corresponds to the standard treatment (Adzick et al. 2011). The primary outcome was defined as a composite binary endpoint of fetal or neonatal death or the need for a cerebrospinal fluid shunt placement among children who survived. The study was planned to detect a decrease in the composite endpoint from 0.85 in the standard treatment group to 0.57 in the intervention group at a one-sided significance level of 0.025 with a power of 0.99. This results in a required total number of 200 patients.

A number of 78 patients were allocated to the prenatal surgery group compared to 80 patients in the postnatal surgery group. The composite endpoint was observed in 53 patients in the prenatal surgery group (corresponding to rate of 0.679) and in 78 patients in the postnatal surgery group (0.975) which correspond to a relevant treatment effect favoring prenatal surgery. However, most of the events were observed for the shunt component (51 events corresponding to an event rate of 0.654 in the prenatal surgery group; 78 events corresponding to an event rate of 0.679 in the postnatal surgery group), whereas for fetal or neonatal death only 2 events were observed, both in the prenatal surgery group. Thus, there is a small but opposite effect for the harmful component death. It can be discussed if this small negative effect is still acceptable or not, but the results illustrate that the interpretation of the composite effect alone can be difficult when the component effects point in adverse directions.

19.2.1 Correlation-Adjusted Local Significance Levels

As the power of the MOMS Trial given by 0.99 is very high, additional confirmatory tests for the components might be included in the formal efficacy claim within a union-intersection test without the need to increase the sample size. If the composite is tested at the reduced level of 0.02, the power to test exclusively the composite at this level still amounts to 0.99, so the power loss is extremely small.

As stated above, the anticipated composite rate of 0.85 in the standard treatment group was based on the assumption of a 0.8 shunt rate and a 0.05 death rate. The assumed event rates for the individual components are thus of very different magnitude. Therefore, it might be appropriate to choose different local significance levels for the components where a higher level is assigned to the more promising endpoint corresponding to the shunt placement. Under the assumption that all null hypotheses are true and the event rates are given as specified above, the correlations between the composite and the individual components can be calculated as introduced in Chap. 8. The correlations are given as

$$\begin{aligned} Corr\,(T_{CE}, T_{Death}) &= 0.096, \\ Corr\,(T_{CE}, T_{Shunt}) &= 0.84, \\ Corr\,(T_{Death}, T_{Shunt}) &= -0.459. \end{aligned}$$

The local levels for the components can be chosen as

$$\alpha^{Death} = \alpha^{Shunt} = 0.0045.$$

However, in this trial, the event rates for the individual components are of very different magnitude. An unexpectedly large effect is therefore much more likely to be observed in the shunt endpoint. Therefore, it might be more appropriate to choose the local levels for the components differently. An alternative choice of the local levels is given by

$$\alpha^{Death} = 0.001,\ \alpha^{Shunt} = 0.01.$$

With this allocation, a large effect in the shunt components is more likely to be detected, whereas an overwhelming effect in the death component seems unrealistic. Note that the above setting of the local significance levels highly profits from the underlying correlation structure, as the sum of the local levels given as

$$\alpha^{CE} + \alpha^{Death} + \alpha^{Shunt} = 0.02 + 0.001 + 0.01 = 0.031$$

considerably exceeds the one-sided family-wise level of 0.025.

This exemplary illustration of correlation-adjusted local significance levels was originally published by Rauch and Kieser (2012).

References

Adzick, N. S., Thom, E. A., Spong, C. Y., Brock, J. W., Burrows, P. K., Johnson, M. P., et al. (2011). A randomized trial of prenatal versus postnatal repair of myelomeningocele. *New England Journal of Medicine, 364*, 993–1004.

Rauch, G., & Kieser, M. (2012). Multiplicity adjustment for composite binary endpoints. *Methods of Information in Medicine, 51*, 309–317.

Rauch, G., & Kieser, M. (2013). An expected power approach for the assessment of composite endpoints and their components. *Computational Statistics & Data Analysis, 60*, 111–122.

Yuksel, N., Majumdar, S. R., Biggs, C., & Tsuyuki, R. T. (2010). Community pharmacist-initiated screening program for osteoporosis: Randomized controlled trial. *Osteoporosis International, 21*, 391–398.

Chapter 20
Clinical Trial Examples with (Composite) Time-to-Event Endpoints

Similar to the previous chapter, we discuss in this chapter several original clinical trials for time-to-event composite endpoints along with their designs, results, and possible limitations. The focus is again to propose and apply alternative planning and analysis strategies which might help to improve the performance of the design or to ease interpretation of results.

20.1 The OMEGA Trial

The OMEGA Trial was a randomized, placebo-controlled, double-blind, multicenter trial to evaluate the effect of omega-3-acid ethyl esters-90 on time to sudden cardiac death in survivors of acute myocardial infarction (Rauch et al. 2010). The primary endpoint was "time to sudden cardiac death" defined as unexpected death resulting from heart disease, occurring within 1 h of the first symptoms or unwitnessed, overnight. The primary endpoint was thus a time-to-event endpoint but not a composite endpoint. However, "time to death from any cause" was defined as a secondary endpoint, where the different causes of death ("sudden cardiac death," "non-sudden cardiac death," or "non-cardiac death") can be interpreted as different types of event. Patients were randomized in a 1:1 allocation to receive either the intervention or placebo. Recruitment duration was not reported, and the minimal patients' follow-up duration was chosen to be 1 year. The sample size calculation was based on detecting a relative reduction of 0.44 in the 1-year cumulative event rate of sudden cardiac death from 0.035 in the placebo group to 0.019 in the OMEGA group with 0.8 power at a global one-sided significance level of 0.025. Note that for the original sample size calculation, the primary composite endpoint was considered as a binary outcome and not as a time-to-event endpoint, which clearly would have been more appropriate here. Incorporating a drop-out rate of 0.088, the planned total sample size was 3800.

The results of the OMEGA Trial were not provided by means of the corresponding hazard ratios, which would have been the appropriate effect measure, but instead by means of the absolute and relative frequencies, thus ignoring the censoring

G. Rauch et al., *Planning and Analyzing Clinical Trials with Composite Endpoints*, Springer Series in Pharmaceutical Statistics,
https://doi.org/10.1007/978-3-319-73770-6_20

distribution and issues of competing risks. Therefore, the published results should be interpreted with care. However, the published results are sufficient to discuss general problems regarding the choice of the primary endpoint. A total of 1919 patient was allocated to the OMEGA group compared to 1885 in the placebo group. The primary endpoint was reached by exactly the same number of 28 patients in both groups (observed rates 0.015) within the follow-up period of 1 year which corresponds to an absence of any treatment effect. However, as the number of observed events and thus the resulting event rates were much smaller than expected in the planning stage, the study was seriously underpowered. This might be a situation where a composite endpoint combining other events of interest with the event "sudden cardiac death" might have been helpful.

20.1.1 Single Endpoint versus Composite Endpoint

The original primary endpoint of the OMEGA Trial was "time to sudden cardiac death". However, the total number of events in the primary endpoint was relatively low. This can be explained by the fact that the current standard treatments for survivors of acute myocardial infarction are already very effective, and thus the potential improvement of new additional treatments can only be small. Therefore, a composite endpoint combining different events of interest might have been more appropriate in this case. For example, "time to death from any cause", which can be interpreted as a composite endpoint combining different causes of death, could be defined as an alternative primary endpoint.

As "time to death from any cause" was considered as a secondary endpoint within the OMEGA Trial, it is possible to directly compare the results of the original individual time-to-event endpoint and the composite endpoint. A reanalysis of the original clinical trial data revealed the results displayed in Table 20.1. Note that although throughout the book one-sided test hypotheses and corresponding p-values are considered, the two-sided p-values of the corresponding logrank test are additionally provided here for the sake of completeness.

The effect for the composite endpoint in terms of the hazard ratio is given by 0.751 which is considerably larger than the cause-specific hazard ratio for sudden cardiac death given by 0.975. This enlargement in the effect illustrates the possible benefit of a composite endpoint over a single time-to-event endpoint. However, it is important to note that the effects of both endpoints point in the adverse direction thus favoring the control, although the two-sided p-value of the composite misses

Table 20.1 Treatment effects for different endpoints of the OMEGA Trial

	Hazard ratio control/omega	One-sided p-value	Two-sided p-value
Sudden cardiac death	0.975	0.538	0.925
Death from any cause	0.751	0.969	0.062

significance, which is reported here to underlying the magnitude of the adverse effect. This exemplary illustration was similarly published by Rauch et al. (2014b).

20.1.2 *Descriptive Evaluation of the Components*

As discussed in the previous section, the OMEGA Trial would have profited from a composite endpoint given by "time to death from any cause" which combines different causes of death. In this case, one of the components is given as "sudden cardiac death" (SCD) which was the original primary endpoint of the OMEGA Trial. The other causes of death which define the remaining components of the composite are given by "non-sudden cardiac death" (NSCD) and "non-cardiac death" (NCD). Following the guideline recommendations, these components should be analyzed separately to provide a complete picture of the composite endpoint. Table 20.2 gives a descriptive analysis of these components by providing the empirical cumulative incidence functions evaluated at 12 months along with the estimated cause-specific hazard ratios and corresponding 95 % confidence intervals. Again, the results are based on a reanalysis of the original clinical trial data as published by Rauch et al. (2014b).

The all-cause hazard ratio for the composite endpoint as well as all cause-specific hazard ratios for the components show an effect in favor of the control group. The negative treatment effects are also supported by the empirical cumulative incidence functions evaluated at 12 months. However, the magnitude of the effects is quite different. While in terms of the cause-specific hazard ratio the negative effect is particularly large for the component "non-sudden cardiac death" where the cause-specific hazard ratio is 0.582, the difference between the OMEGA and the control group in the empirical cumulative incidence functions evaluated at 12 months is only $0.02 - 0.012 = 0.008$. This indicates that the empirical cumulative incidence functions alone may provide misleading results on the component treatment effects.

In an extreme scenario, the comparison of the empirical cumulative incidence functions could even show a higher event rate in the intervention group for a specific component, whereas the corresponding cause-specific hazard ratio favors the intervention. To illustrate this, we additionally provide a descriptive analysis

Table 20.2 Descriptive analysis of composite and the components for the OMEGA Trial

	CIF omega	CIF control	Hazard ratio control/omega
Death from any cause	0.046 (0.043; 0.05)	0.037 (0.035; 0.04)	0.751 (0.555; 1.016)
Sudden cardiac death	0.015 (0.014; 0.017)	0.015 (0.014; 0.017)	0.975 (0.58; 1.639)
Non-sudden cardiac death	0.020 (0.019; 0.022)	0.012 (0.011; 0.014)	0.582 (0.349; 0.972)
Non-cardiac death	0.011 (0.010; 0.012)	0.009 (0.008; 0.01)	0.968 (0.532; 1.76)

CIF: empirical cumulative incidence function at 12 months; CIFs and hazard ratios provided with 95 % confidence intervals

Table 20.3 Descriptive analysis of the composite and the components for the simulated data motivated by the OMEGA Trial

	CIF omega	CIF control	Hazard ratio control/omega
Death from any cause	0.910 (0.865; 0.945)	0.989 (0.963; 0.998)	2.354 (1.900; 2.918)
Sudden cardiac death	0.675 (0.610; 0.739)	0.366 (0.302; 0.439)	1.174 (0.868; 1.586)
Non-sudden cardiac death	0.075 (0.046; 0.121)	0.539 (0.470; 0.611)	16.390 (9.356; 28.713)
Non-cardiac death	0.160 (0.116; 0.219)	0.083 (0.052; 0.133)	1.071 (0.572; 2.004)

CIF: empirical cumulative incidence function at 3 months; CIFs and hazard ratios provided with 95 % confidence intervals

of a simulated data set with the same rounded sample size of 3850 as given for the OMEGA Trial. To simulate the event times related to the three different causes of death, exponentially distributed event times were generated with the following parameters:

$$\theta_{CE} = \frac{1.85}{0.85} = 2.176,$$

$$\theta_{SCD} = \frac{0.75}{0.5} = 1.500,$$

$$\theta_{NSCD} = \frac{0.833}{0.083} = 10.036,$$

$$\theta_{NCD} = \frac{0.167}{0.167} = 1.000.$$

The time scale for the above hazard assumptions is based on months. Note that this extreme scenario is chosen in order to demonstrate the differences between the cause-specific hazard ratios and the cumulative incidence functions. Table 20.3 gives a descriptive analysis of these components by providing the empirical cumulative incidence functions evaluated at 3 months along with the estimated cause-specific hazard ratios and corresponding 95 % confidence intervals. In addition, Fig. 20.1 displays the empirical cumulative incidence functions over time for the composite and the components. In contrast to the original OMEGA Trial data, event times were already censored after 3 months, because the simulated data are based on much higher cause-specific hazards and yielded therefore higher event rates.

It can be seen that the estimated hazard ratios of the simulated data set are reasonably close to the theoretical hazard ratios provided above. In terms of the all-cause and cause-specific hazard ratios, the intervention tends to have a positive effect on the composite endpoint and all components. However, if the cumulative incidence functions at 3 months are considered, the results are quite different. For "sudden cardiac death," the empirical cumulative incidence function at 3 months is considerably larger in the OMEGA group than in the control group. The same holds true for the component "non-cardiac death". However, these findings cannot be interpreted as negative treatment effects. The low event rates for "sudden cardiac

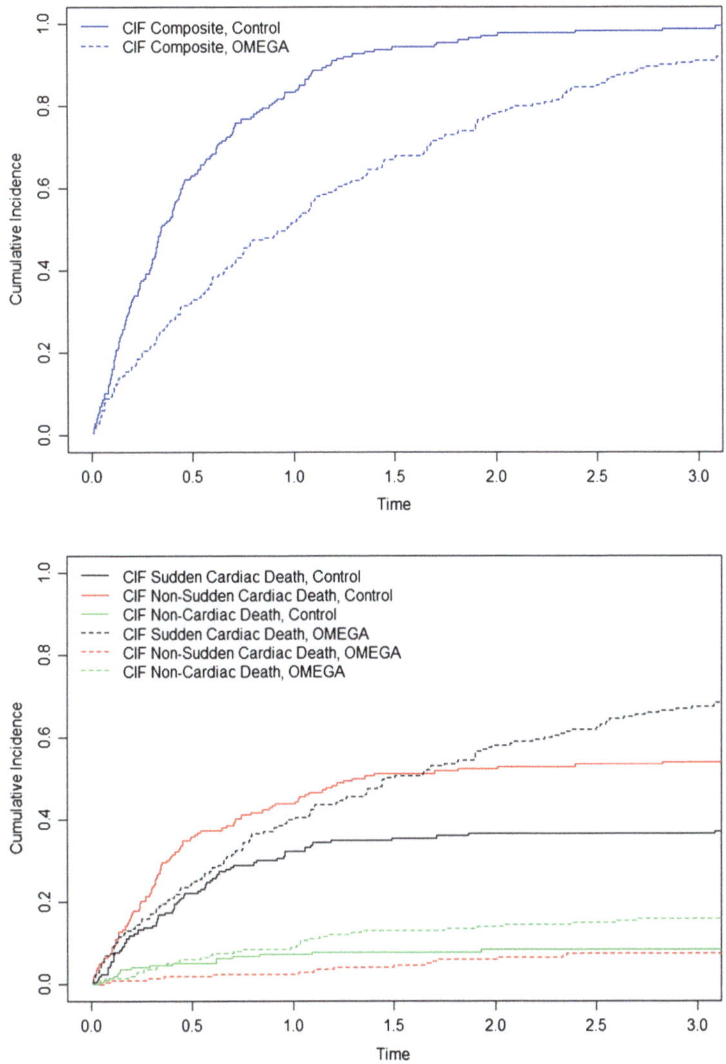

Fig. 20.1 Empirical cumulative incidence functions over time for the composite and the components for the simulated data motivated by the OMEGA Trial

death" and "non-cardiac death" in the control group can be explained by the influence of the competing component "non-sudden cardiac death" which shows a high estimated event probability in the control group. Therefore, the event rates for "sudden cardiac death" and "non-cardiac death" in the control group are only low because many patients died of non-sudden cardiac causes before. In contrast, the related cause-specific hazard ratios are not influenced by competing events. A simulation study illustrating a similar scenario was published by Rauch et al. (2014b).

20.1.3 Incorporating Binding Stopping for Futility

As shown above, the original primary endpoint of the OMEGA Trial failed significance by far. Therefore, this trial would have highly profited from a group-sequential design with a binding futility stopping rule at the interim stage. For the sake of illustration, let us consider a group-sequential design with one interim analysis after a fraction of 0.3 of the total number of events was observed. We further assume exponentially distributed event times. With the given sample size of 3800, the expected total number of events under the original planning assumptions is 250. As the recruitment duration of the OMEGA Trial was not reported, we assume for illustrative purposes a recruitment duration of 3 years. The required total number of patients yielding $0.3 \cdot 250 = 75$ events is expected to be observed after inclusion of 2660 patients, which corresponds to a fraction of 0.525 of the total sample size. The local significance levels are determined according to Pocock and thus amount to

$$\alpha_{1,0.025} = \alpha_{1+2,0.025} = 0.0147$$

for a global one-sided significance level of 0.025.

The event rates assumed in the planning stage to calculate the original sample size translate to a hazard ratio for sudden cardiac death (SCD) of

$$\theta_{SCD} = \frac{0.00297}{0.0016} = 1.856,$$

which thus corresponds to the assumed effect η under the alternative hypothesis using the notations introduced in Definitions 6.4.1 and 6.4.2 in Chap. 6 in Part II of this book. Note that the hazards specified above are based on a time scale using months rather than years. Remember that the original sample size calculation of the OMEGA Trial was performed using a test for comparing rates rather than by means of the logrank test, which would have been more appropriate in this setting. The power based on the logrank test for the above parameter setting is given by 0.998 and thus much higher than the above-stated value of 0.8.

In order to define the OCS futility boundary for the interim stage according to Definition 6.4.2, the maximal admissible overall power loss β_{loss} must be prespecified. In view of the high power under the original planning assumptions given by 0.998, we will assume that a power loss of at most $\beta_{loss} = 0.2$ is still acceptable. Moreover, the maximal admissible rate for wrongly stopping for futility is set to $\pi_{wrong} = 0.025$. An effect in terms of the hazard ratio given as $\eta_{stop} = 1.3$ is considered as the largest effect for which stopping the study for futility would still be considered as "correct." For these settings, the OCS futility boundary and the corresponding quantile are given as

$$\alpha_{opt} = 0.24, \; z_{1-\alpha_{opt}} = 0.706.$$

With this binding futility bound, the probability to correctly stop for futility under the effect $\eta_{stop} = 1.3$ amounts to 0.305. The probability of wrongly stopping for futility is given by 0.024. The power with stopping for futility amounts to 0.974. Thus, the actual power loss $0.998-0.974 = 0.024$ is much smaller than the maximal acceptable power loss of 0.2. However, for a smaller futility boundary which would imply a higher loss in power, the maximal admissible rate for wrongly stopping for futility would no longer be controlled by $\pi_{wrong} = 0.025$.

If we assume that at the interim stage the expected effect in terms of the hazard ratio is equivalent to the one observed in the original trial given by 0.975 and if we moreover assume that 20 events in total are observed at interim, then the observed value of the test statistic at interim is approximately given as

$$w_{SCD}^{1} \approx \ln(\theta_{SCD}) \sqrt{\frac{n_I \cdot n_C}{n_I + n_C} \cdot \pi_{CE}} = \ln(0.975) \sqrt{\frac{10}{2}}$$

$$= \ln(0.975) \sqrt{\frac{10}{2}} = -0.057$$

compare (5.21) in Chap. 5 of Part II. As

$$w_{SCD}^{1} = -0.057 \leq z_{1-\alpha_{opt}} = 0.706,$$

the trial would have been stopped for futility and a total of 1140 patients would have been saved. This illustrates the benefit of incorporating stopping for futility.

20.2 The RENAAL Trial

The RENAAL Trial is a typical example for showing how difficult the interpretation of a composite endpoint as primary outcome variable can be. The study was designed as a randomized, double-blind, placebo-controlled clinical trial with one interim analysis to assess whether the angiotensin-II-receptor antagonist losartan shows a therapeutic benefit for nephropathy patients with type 2 diabetes (Brenner et al. 2001, 2000). The primary endpoint was a composite time-to-first-event endpoint where the event types were given as "death," "end-stage renal disease," or "doubling in the baseline serum creatinine concentration." Patients were randomized in a 1:1 allocation to receive either losartan or placebo. Recruitment duration was planned to require 2 years, and the minimal patients' follow-up duration was chosen to be 3.5 years. The sample size calculation was based on detecting a relative risk reduction of 0.2 in the 5-year cumulative event rate of the composite endpoint from 0.58 in the placebo group to 0.464 in the losartan group with 0.95 power at a global one-sided significance level of 0.025, where the adjusted local significance level of the final analysis was 0.024. Assuming exponentially distributed event times, the above effect corresponds to a hazard ratio of 1.392 (control versus intervention).

The total sample size of 1513 patients included a considerable number of patients due to a recruitment overrun.

Although hazard ratios would have been the appropriate effect measures in this context, the results of the RENAAL were provided by means of the absolute and relative frequencies which ignores the problem of censoring and competing risks. As a consequence, the published results should be interpreted with care. Nevertheless, these results are used to discuss general issues with respect to the composite primary endpoint. A number of 975 patients was recruited to the losartan group and 984 patients to the placebo group. The primary composite endpoint was reached by 327 patients in the losartan group (0.435) as compared to 359 patients in the placebo group (0.471) which corresponds to a relevant positive treatment effect. A detailed look into the single component effects reveals a relevantly lower risk in the losartan group for the components "doubling in the baseline serum creatinine concentration" (162 events corresponding to an event rate of 0.216 in the losartan group versus 198 corresponding to a rate of 0.260 in the control group) and "end-stage renal disease" (147 events corresponding to an event rate of 0.196 in the losartan group versus 194 events corresponding to an event rate of 0.255 in the control group), whereas in the particular harmful component "death," a small adverse effect was observed (158 events corresponding to an event rate of 0.210 in the losartan group versus 155 events corresponding to an event rate of 0.203 in the control group). These results illustrate the difficulty in the interpretation of a composite endpoint in case a large effect in a single component masks an adverse effect in another severe component.

20.2.1 Intersection-Union Versus Effect Consistency Approach

As discussed above, a possible shortcoming of the RENAAL Trial is that the composite effect cannot be interpreted easily as the components are of different clinical relevance. In particular, a major adverse effect in an important component like "death" should be ruled out. A possible solution would thus be to define an adequate multiple test procedure including the composite endpoint and the main component given by death.

The most simple idea would be to introduce an additional non-inferiority or superiority test for the endpoint "death" to assess if the intervention is non-inferior or superior to the control with respect to this important component. An intuitive multiple test strategy is thus given by the intersection-union test, compare Chap. 9 of Part III of this book. For the sake of comparison, we investigate different non-inferiority margins for the component death given as $\exp(\delta_{Death}) = \exp(-0.105) = 0.9$ as a strong and $\exp(\delta_{Death}) = \exp(-0.22) = 0.8$ as a more liberal non-inferiority margin. Moreover, the intersection-union test based on the superiority test problem for the component "death" will be also investigated, i.e., $\exp(\delta_{Death}) = \exp(0) = 1$.

An alternative multiple test strategy to assess the composite endpoint and one main component is given by the effect consistency approach presented Chap. 9 of Part III of this book as well. The general idea is to introduce a consistency bound for

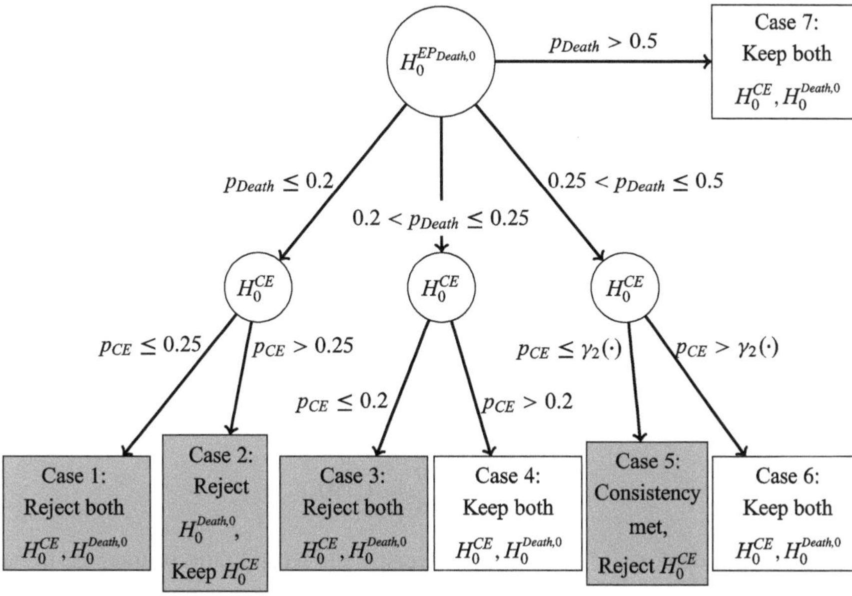

Fig. 20.2 Effect consistency approach for the RENAAL Trial

the main component instead of performing a formal hypothesis test. In this situation, a possible consistency claim for the main component "death" could be that the effect should not point in the wrong direction, which means that the one-sided p-value of the superiority test should not be smaller than the consistency bound given as $\alpha^* = 0.5$. For a one-sided global significance level of $\alpha = 0.025$, we further choose $\alpha_I = \alpha_{II} = 0.2$. Figure 20.2, similarly given in Rauch et al. (2014c), illustrates the effect consistency approach for the RENAAL Trial with the above specified parameters. In the following, the two alternative multiple test procedures (intersection-union test and effect consistency approach) will be compared in terms of power and gain in information for the specific situation of the RENAAL Trial.

The RENAAL Trial was planned to detect a 5-year reduction in the cumulative composite event probability from 0.58 in the placebo group to 0.464 in the treatment group (0.2 relative reduction). Assuming exponentially distributed event times, the above effect in the composite event rates corresponds to a hazard ratio of

$$\theta_{CE} = \frac{0.174}{0.125} = 1.392.$$

For the simulations, an additional assumption on the effect size for the component "death" is required. Thereby, it will be assumed that the cause-specific hazards for death correspond to one third of the all-cause hazards corresponding to the composite endpoint resulting in an equivalent hazard ratio of

$$\theta_{Death} = \frac{0.058}{0.04166} = 1.392.$$

These assumptions are referred as the *assumed effects scenario* in the following. Again, the hazards specified above are based on a time scale using months.

Looking at the results of the RENAAL Trial, it can be seen that the observed effects were quite different from the anticipated ones. Thereby, as stated above, the results of the RENAAL Trial were not reported as hazard ratios and therefore cannot directly be presented to compare them to the original planning assumptions. However, based on the published results in terms of absolute and relative event frequencies, the observed composite effect in terms of the hazard ratio can be reasonably assumed to be given by

$$\theta_{CE} = \frac{0.156}{0.126} = 1.190.$$

For the main component "death", the cause-specific hazard ratio is assumed to be given by

$$\theta_{Death} = \frac{0.020}{0.021} = 0.952,$$

which corresponds to a small adverse effect as more deaths were observed under the intervention. The above effects will be referred as the *observed effects scenario* in the following. A simulation study based on 10,000 runs using the software R was performed to assess the power of the different multiple test procedures for both scenarios given by the assumed and by the observed effects. Thereby, the original sample size of 1513 was rounded to 750 patients per group (1500 in total). The results are displayed in Table 20.4.

Table 20.4 Power comparisons for different efficacy claims

		Simulated power to reject ...				
			$H_0^{CE} \bigcap H_0^{Death,\delta_{Death}}$	H_0^{CE} (Cases 1, 3, 5)	Cases 1, 2, 3, 5	
Scenario	δ_{Death}	H_0^{CE}	with IUT	with ECA	with ECA	Case 5
Assumed effects	0.00	0.993	0.723	0.976	0.978	0.253
	−0.105	0.993	0.916	0.976	0.978	0.253
	−0.22	0.993	0.985	0.976	0.978	0.253
Observed effects	0.00	0.808	0.014	0.312	0.312	0.298
	−0.105	0.808	0.046	0.312	0.312	0.298
	−0.22	0.808	0.131	0.312	0.312	0.298

IUT: intersection-union test, *ECA*: effect consistency approach

Column 3 of Table 20.4 shows the simulated power to test exclusively the composite endpoint, which is relatively large as the sample size included a considerable recruitment overrun. In Column 4, the power to reject both the superiority or non-inferiority null hypothesis for the main component and the superiority null hypothesis for the composite endpoint by means of the intersection-union test is provided. The power loss of the intersection-union test compared to the single endpoint problem assessing the composite endpoint alone strongly depends on the underlying effects. For the assumed effects scenario, the effect in the component death is reasonably large, and therefore the power loss is limited. In contrast, for the observed effects scenario, the power loss is huge as the death component shows a slight adverse effect.

Columns 5–7 show the results for the effect consistency approach. The correlation was bootstrapped from the control group sample with the algorithm given by Rauch et al. (2014c) which is also provided as an R code in Chap. 11 in Part III of this book. For the assumed effects scenario, the correlation is approximately given by 0.706. For the observed effects scenario, the correlation is 0.576. Note that the probability to reject the composite null hypothesis given in Column 8 is equal to the joint probability for Cases 1, 3, and 5 (compare Fig. 20.2). Therefore, this probability differs only slightly from the power given in Column 6 which includes the probability for Case 2. As displayed in Fig. 20.2, Case 2 corresponds to the probability to obtain a significant effect in the main component but not in the composite endpoint. For the considered scenarios, this probability is extremely small. As the new aspect of the effect consistency approach is that it is sufficient to show superiority with respect to the composite endpoint and simultaneously a consistent effect in the main component (which corresponds to Case 5), the probability for Case 5 is provided in the last Column 7. This quantifies the gain in power compared to a formal hypothesis test of the main component. The effect consistency approach shows a better power performance than the intersection-union test based on a superiority test or a non-inferiority test with margin $\delta_{Death} = -0.105$ for the main component. For the observed effects scenario, this also holds true for a non-inferiority test with margin $\delta_{Death} = -0.2$ for the main component. However, a formal hypothesis test for the main component clearly provides more information than the consistency claim. Therefore, it cannot generally be stated which method should be preferred. The choice of the method depends on the available resources and on a careful consideration about the required additional information which is necessary for a valid interpretation of the specific trial.

20.2.2 Weighted All-Cause Hazard Ratio

As explored above, the composite effect of the RENAAL Trial may not be easily interpreted as the components are of different clinical relevance. Instead of formulating a multiple test problem as proposed in the previous section, an alternative may be to use a weighted effect measure which assigns higher weights to clinically more relevant components. In this section, the weighted all-cause hazard ratio introduced in Chap. 13 in Part IV will be applied to the RENAAL Trial.

The primary composite endpoint of the RENAAL Trial included the components "death," "end-stage renal disease," and "doubling in the baseline serum creatinine concentration." Clearly, death defines the most relevant component, followed by end-stage renal disease which also corresponds to a very severe outcome. Doubling of baseline serum creatine, however, is a rather soft endpoint as the severeness of this event clearly depends on the baseline serum creatine value. A doubling of a very low baseline value might be harmless, whereas a doubling in a high baseline value can be most relevant. Therefore, this component is not a good measure to assess the treatment effect. Based on these relevance considerations, we choose the following weights:

$$w_{Death} = 0.5, \ w_{EndStage} = 0.4, \ w_{Doubling} = 0.1.$$

To check the robustness of the results under small changes of the weights, the following additional weight scenarios will be considered:

$$w_{Death} = 0.6, \ w_{EndStage} = 0.3, \ w_{Doubling} = 0.1,$$

$$w_{Death} = 0.5, \ w_{EndStage} = 0.3, \ w_{Doubling} = 0.2.$$

Moreover, the situation of equal weights across components

$$w_{Death} = w_{EndStage} = w_{Doubling} = 1$$

will be investigated to compare the results to the standard all-cause hazard ratio. Again, we assume exponentially distributed event times for the composite and all components implying that the weighted all-cause hazard ratio is constant in time. For data simulation, we again consider two scenarios corresponding to the assumed and the observed effects. The following hazards on a monthly time scale are used to simulate the anticipated effects:

$$\theta_{CE} = \frac{0.174}{0.125} = 1.392,$$

$$\theta_{Death} = \frac{0.058}{0.04166} = 1.392,$$

$$\theta_{EndStage} = \frac{0.058}{0.04166} = 1.392,$$

$$\theta_{Doubling} = \frac{0.058}{0.04166} = 1.392.$$

To simulate the observed effects, the following hazards are assumed which are motivated by the original results of the RENAAL Trial:

$$\theta_{CE} = \frac{0.156}{0.126} = 1.238,$$
$$\theta_{Death} = \frac{0.020}{0.021} = 0.952,$$
$$\theta_{EndStage} = \frac{0.071}{0.054} = 1.315,$$
$$\theta_{Doubling} = \frac{0.065}{0.051} = 1.275.$$

A simulation study based on 10,000 runs using the software `R` was performed to assess the performance properties of the weighted all-cause ratio and the related permutation test for the assumed and the observed effects scenario. Thereby, again a sample size of 750 patients per group was assumed. For each sample, the corresponding hazard estimates within the two groups were obtained by fitting Weibull models with scale parameters fixed to 1 and thus corresponding to exponential models. The plug in estimator $\widehat{\theta}_{CE}^{w}$ was obtained by substituting the resulting hazard estimates. The permutation test was performed by randomly reassigning the group affiliations to the original complete data set and subsequently calculating the point estimator $\widehat{\theta}_{CE}^{w}$ as described above. We performed 1000 repetitions of permutations to obtain the null distribution of the estimator $\widehat{\theta}_{CE}^{w}$. The simulation results are provided in Table 20.5.

For the assumed effects scenario, it can be observed that the estimated treatment effects in terms of the all-cause hazard ratio and the weighted all-cause mean hazard ratio are very similar for all considered weights. This is due to the fact that the effects in the individual components were assumed to be equivalent. The power values of the logrank test and the permutation test are comparable as well.

Table 20.5 Simulated weighted all-cause hazard ratio and power based on the RENAAL Trial

Scenario	w_{Death}	$w_{EndStage}$	$w_{Doubling}$	$\widehat{\theta}_{CE}^{w}$ (Power)	$\widehat{\theta}_{CE}$ (Power)
Assumed effects	1.0	1.0	1.0	1.387 (0.994)	1.387 (0.994)
	0.5	0.4	0.1	1.390 (0.979)	1.389 (0.994)
	0.6	0.3	0.1	1.388 (0.968)	1.390 (0.995)
	0.5	0.3	0.2	1.389 (0.985)	1.389 (0.993)
Observed effects	1.0	1.0	1.0	1.235 (0.810)	1.235 (0.808)
	0.5	0.4	0.1	1.201 (0.585)	1.234 (0.806)
	0.6	0.3	0.1	1.170 (0.433)	1.236 (0.809)
	0.5	0.3	0.2	1.198 (0.631)	1.235 (0.806)

$\widehat{\theta}_{CE}^{w}$: estimated weighted all-cause hazard ratio and related power of the permutation test
$\widehat{\theta}_{CE}$: estimated all-cause hazard ratio and related power of the logrank test

From the observed effects scenario, it can be seen that the weighted all-cause hazard ratio is reasonably robust against small changes in the weights. As the highest weight is assigned to the death component which shows a slight adverse effect, the weighted all-cause hazard ratio is smaller than the all-cause hazard ratio. As a consequence, the power is also lower. As the intention of the weighted approach is to improve the interpretation of the composite effect measure, it is desirable that the weighted effect is shrunken when the most relevant component corresponds to an adverse effect. In this application, the weighted all-cause hazard ratio therefore better reflects the net effect of the treatment.

Another clinical trial application illustrating the performance of the all-cause hazard ratio and the related permutation test was published by Rauch et al. (2018).

20.3 The DREAM Trial

The DREAM Trial was a randomized, double-blind, placebo-controlled study to demonstrate that the drug rosiglitazone can prevent high-risk patients from developing type 2 diabetes (The DREAM Trial Investigators 2006). The primary endpoint of the trial was a time-to-first-event endpoint composed of "death" or "newly developed diabetes." A total of 5269 patients were randomized in a 1:1 allocation ratio. The minimal patients' follow-up duration was chosen to be 2 years. The study was planned to provide a power of 0.9 to detect a relative reduction of 0.22 from a rate of 0.045 at 1 year in the placebo group to a rate of 0.035 in the intervention group at a one-sided significance level of 0.025.

The results showed a clear benefit of the intervention. The all-cause hazard ratio for the composite endpoint was given by 2.5 (control versus intervention). However, the effects of the individual components were of very different magnitude. The hazard ratio for death was 1.10, whereas for newly developed diabetes it was given by 2.63. Thus, the composite effect is mainly driven by the less severe endpoint "newly developed diabetes" which may question the composite effect as a meaningful measure to assess the treatment effect.

20.3.1 *Properties of the Proportion in Favor of Treatment*

As presented above, the primary endpoint of the DREAM Trial was a composite of death and newly developed diabetes. The observed effect in the composite endpoint was given by $\widehat{\theta}_{CE} = 2.5$, but this effect is mainly driven by a large effect in the less relevant component "newly developed diabetes". For the patient, the death component clearly is the more relevant endpoint. Therefore, a weighted effect measure might seem more appropriate here. In the following, we will investigate whether the proportion in favor of treatment for prioritized outcomes assigning first priority to the death component might be a better outcome in this situation.

As the original data of the above trial were not available, we assumed the event times to be exponentially distributed. We further assumed a constant follow-up time of 2 years for each patient. It is not directly possible to determine the individual hazards from the results reported by the DREAM Trial Investigators (2006). As death corresponds to a fatal event, the components obviously define competing risks, and the individual hazard is not identifiable. However, as the DREAM Trial is only used for the sake of illustration, it will be assumed in the following that the individual hazards can be reasonably estimated and that the competing risk scenario has no major influence. Considering the published event rates and hazard ratios for the individual endpoints can help to get an idea of the magnitude of the underlying individual hazards. For the sake of illustration, we therefore chose the individual hazards on a monthly time scale to be given by

$$\widetilde{\theta}_{Death} = \frac{0.006}{0.00545} = 1.1,$$

$$\widetilde{\theta}_{Diab} = \frac{0.143}{0.055} = 2.6.$$

Recall that the individual hazard ratio $\widetilde{\theta}$ differs from the cause-specific hazard ratio θ by ignoring events of other type. The theoretical proportion in favor of treatment for a constant follow-up time f is given as

$$\begin{aligned}\Omega = &\frac{\widetilde{\theta}_{Death} - 1}{\widetilde{\theta}_{Death} + 1} \cdot \underbrace{\left(1 - e^{-(\widetilde{\lambda}^I_{Death} + \widetilde{\lambda}^C_{Death}) \cdot f}\right)}_{=\vartheta^{times}_{Death}(f)} \\ &+ \frac{\widetilde{\theta}_{Diab} - 1}{\widetilde{\theta}_{Diab} + 1} \cdot \underbrace{\left(1 - e^{-(\widetilde{\lambda}^I_{Diab} + \widetilde{\lambda}^C_{Diab}) \cdot f}\right) \cdot e^{-(\widetilde{\lambda}^I_{Death} + \widetilde{\lambda}^C_{Death}) \cdot f}}_{=\vartheta^{times}_{Diab}(f))}.\end{aligned}$$

For the parameter constellations given above, the proportion in favor of treatment is thus given by

$$\Omega = 0.143,$$

which corresponds to a considerably smaller effect than the transformed all-cause hazard ratio given by

$$\frac{\widehat{\theta}_{CE} - 1}{\widehat{\theta}_{CE} + 1} = 0.429.$$

Looking at the transformed individual hazard ratios given as

$$\frac{\widehat{\theta}_{Death} - 1}{\widehat{\theta}_{Death} + 1} = 0.048,$$

$$\frac{\widehat{\theta}_{Diab} - 1}{\widehat{\theta}_{Diab} + 1} = 0.444$$

it becomes obvious that the proportion in favor of treatment is driven toward the transformed individual hazard ratio for death. The proportion in favor of treatment, however, strongly depends on the censoring time, i.e., on the study duration. The impact of this problem becomes more illustrative if we consider the proportion in favor of treatment as a function of the constant follow-up time f as presented in Fig. 20.3.

It can be seen that the slope of $\Omega(f)$ is huge within the interval [0; 10]. This means that a small change in the follow-up time will have a huge impact on Ω. In Fig. 20.3, $\Omega(2) = 0.143$ and $\Omega(3) = 0.194$ are displayed as references lines for illustration. When the follow-up time is close to 0, say $f = 0.5$, then $\Omega(0.5) = 0.042$, where the weights are given as $\vartheta^{times}_{Death}(0.5) = 0.034$ and $\vartheta^{times}_{Diab}(0.5) = 0.433$. As $\vartheta^{times}_{Death}(0.5)$ is much larger than $\vartheta^{times}_{Diab}(0.5)$, the second

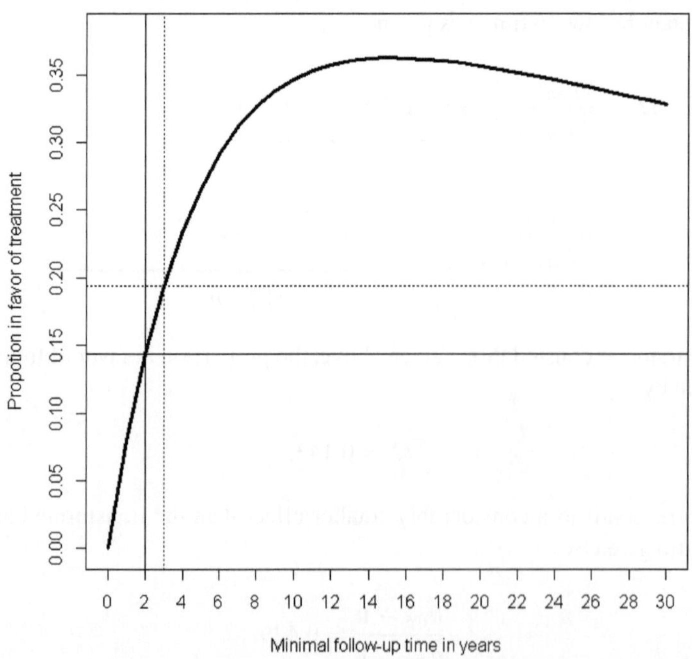

Fig. 20.3 Proportion in favor of treatment for 2 and 3 years of constant follow-up for the parameter constellation of the DREAM Trial

component dominates the effect. Note, however, that the sum of the weights is much smaller than 1, and thus the whole effect is shrunken. With increasing f, on the one hand, the sum of the weights increases as well so that the effect gets larger. On the other hand, the impact of the first components gets larger with increasing f, and as the effect for death is smaller than for the diabetes component, the combined effect measures decrease. For this reason, it is plausible that the univariate effect measure increases with f until a certain time point where the impact of the weight sum cancels out with the impact of the larger weight for the first component. Therefore, the function $\Delta(f)$ is increasing until $f = 15$ and decreases after this time point. From these considerations, it can be seen that the proportion in favor of treatment is difficult to interpret as the magnitude of the effect highly depends on the censoring distribution and is influenced by a number of factors related to the follow-up duration. This clinical trial illustration was originally published by Rauch et al. (2014a).

20.4 The CAPRICORN Trial

A very illustrative example from cardiology that shows how difficult it can be to determine an adequate primary endpoint in the planning stage is the CAPRICORN Trial. Therein, Dargie (2000) investigated the long-term efficacy of carvedilol regarding morbidity and mortality in patients with left ventricular dysfunction after acute myocardial infarction. In this multicenter, randomized, double-blind, and placebo-controlled trial, patients were randomly assigned to carvedilol or placebo in a 1:1 allocation. The originally planned primary endpoint was "time to death from any cause". The trial was planned to detect a hazard ratio for death of 0.77 with a power of 0.9 for a one-sided significance level of 0.025. Recruitment was planned to continue until 633 target events were observed which was assumed to require 1850 patients in total.

During a masked interim analysis, the data safety monitoring board noted that the overall mortality rate was lower than anticipated and consequently the study was likely to be seriously underpowered. Therefore, it was decided to continue the trial with two primary endpoints of which the first corresponds to the original primary endpoint (all-cause mortality) and the second to a composite time-to-first-event endpoint given as time to death or cardiovascular hospital admission. By adding the composite endpoint, it was intended to increase the number of events and thus augmenting the power of the trial. The new efficacy claim of the trial was to show significance in at least one of the two endpoints with higher preference for the more promising composite endpoint. In terms of the multiple test problem arising from this modified analysis, it was therefore decided to assign a weighted Bonferroni-adjusted local significance level of only 0.005 to the original primary endpoint (all-cause mortality), whereas a much larger local level of 0.045 was assigned to the newly added composite endpoint (time to death or cardiovascular hospital admission). The updated target sample size was based on the new composite time-to-first-event endpoint.

After a mean follow-up of 1.3 years and inclusion of 1959 patients, the target number of events was reached for the new composite endpoint. The results of the CAPRICORN Trial published by the CAPRICORN Investigators (2001) showed a hazard ratio for death of 1.30 (control versus intervention) and a hazard ratio of 1.09 for the new composite endpoint. Neither of the endpoints showed a significant effect. The original primary endpoint "time to death" indeed showed the effect assumed in the planning stage, but the underlying event rate was much smaller leading to a loss in power. The new composite endpoint showed more events, but the treatment effect nevertheless was lower than for the original primary endpoint. The observed results illustrate the problem of the high uncertainty about reasonable treatment effect assumptions in the planning stage. The study design modifications based on the blinded interim results were not appropriate to account for these planning uncertainties.

20.4.1 Adaptive Bonferroni-Holm for Two Candidate Endpoints

As motivated above, the choice between a composite endpoint and a single time-to-event endpoint can be difficult as the underlying treatment effects can be difficult to predict in the planning stage. For this reason, the CAPRICORN Trial is an illustrative example to discuss the union-intersection test for the assessment of two candidate endpoints. In addition, due to the high planning uncertainties, the aim is to apply an adaptive design with the option to recalculate the sample size at interim based on the more promising endpoint. We will therefore investigate the "group-sequential Bonferroni-Holm approach" in combination with adaptive sample size recalculation as presented in Chap. 10 in Part III of this book. The resulting design is therefore referred to as the "adaptive Bonferroni-Holm approach" in the following.

To investigate the performance of this design in the setting of the CAPRICORN Trial, a simulation study based on the original planning parameters is implemented. However, the parameter assumptions given in Dargie (2000) were not sufficient to reproduce the sample size as neither the underlying hazards nor the recruitment and minimal follow-up duration were specified. For the sake of illustration and trying to be as close as possible to the assumptions given in original publications (The CAPRICORN Investigators 2001; Dargie 2000), we therefore assume an accrual time of 24 months, a minimal follow-up duration of 3 months, and underlying hazards for all-cause mortality given by

$$\theta_{Death} = \frac{0.033}{0.0254} = 1.3. \tag{20.1}$$

Under these assumptions, the required total sample size for the fixed design with one endpoint and without interim analysis is given by 1804 patients corresponding to 614 events when calculated with the software ADDPLAN (ICON plc 2014) using Schoenfeld's formula. Thereby, exponentially distributed survival times and uniform patient accrual were assumed.

To implement an adaptive design with sample size recalculation based on two candidate endpoints ("death" and a composite (CE) of "death" and "cardiovascular hospital admission"), an interim analysis is assumed to be performed after approximately 18 months of recruitment as soon as 1368 patients were enrolled. At this time, the information fraction is roughly given by 0.5 under the hazard assumptions specified above. The interim analysis is performed directly after enrollment of the 1368 patients without adding a follow-up period. According to the group-sequential Bonferroni-Holm approach, the study is stopped at interim, if the null hypothesis of the endpoint corresponding to the larger observed effect can be rejected at level $\alpha_{1,\alpha/2}$. If the study cannot be stopped at interim, the sample size is recalculated according to formula (10.8) in Chap. 10 in Part III of this book. If the recalculated sample size is more than 5-times higher than the interim sample size, i.e., the recalculated sample size is larger than 6840, the final sample size is set to this limit. At the final analysis, the null hypothesis of the endpoint corresponding to the larger observed effect is tested at the local level $\alpha_{1+2,\alpha/2}$, and the null hypothesis corresponding to the other endpoint is tested at the local level $\alpha_{1+2,\alpha}$. In this application, we choose the same local significance levels for the interim and for the final analysis which corresponds to the common Pocock adjustment (Pocock 1977). For the global one-sided significance level of $\alpha = 0.025$, this implies that

$$\alpha_{1,\frac{\alpha}{2}} = \alpha_{1+2,\frac{\alpha}{2}} = 0.00716,$$
$$\alpha_{1,\alpha} = \alpha_{1+2,\alpha} = 0.0147.$$

The aim is now to apply the adaptive Bonferroni-Holm approach to simulated study data. Thereby, we investigate two scenarios where the first mimics the data under the assumed effects in the planning stage and the second models the truly observed effects of the CAPRICORN Trial. As the CAPRICORN Trial was originally planned to assess exclusively the "death" endpoint, the assumed effects scenario will model the situation that the effects of both candidate endpoints are equivalent. This means, that the effects for both endpoints are exactly the same so that there are no first events corresponding to cardiovascular hospital admission. Therefore, the hazards on a monthly time scale are given by

$$\theta_{CE} = \frac{0.033}{0.0254} = 1.3,$$
$$\theta_{Death} = \frac{0.033}{0.0254} = 1.3.$$

In the observed effects scenario, the effects are given as

$$\theta_{CE} = \frac{0.0299}{0.0276} = 1.08,$$
$$\theta_{Death} = \frac{0.0104}{0.008} = 1.3.$$

For both scenarios, a total of 10,000 simulation runs were performed.

Table 20.6 provides the performance characteristics of the adaptive Bonferroni-Holm approach with sample size recalculation for the two investigated scenarios. We provide the power for each of the two candidate endpoints separately, the power to show significance in both endpoints ($H_1^{Death} \cap H_1^{CE}$), the power to show significance in at least one of the endpoints ($H_1^{Death} \cup H_1^{CE}$), and the amount of cases the trial continues to the second stage. Moreover, the average sample size and its standard deviation are reported.

In order to provide fair reference values, Table 20.7 reports the sample sizes required for a standard non-adaptive one-stage design with two endpoints where the aim is to demonstrate a significant effect in at least one of them. If both endpoints match to the original planning assumptions of the CAPRICORN trial (assumed effects scenario), the total required sample size to achieve a power of 0.9 is given by 2130 which is thus by 326 patients larger than the corresponding one-stage design for a single primary endpoint. For the observed effects scenario, a total sample size of approximately 6600 patients is needed. Table 20.7 also reports the different power values for these two sample sizes under the assumed and the observed effects scenarios.

Table 20.6 Power values for the adaptive Bonferroni-Holm approach

	Power for...					
Scenario	H_1^{Death}	H_1^{CE}	$H_1^{Death} \cap H_1^{CE}$	$H_1^{Death} \cup H_1^{CE}$	ASS ± SD	Go to stage 2
Assumed effects	0.966	0.966	0.966	0.966	1908 ± 1121	0.316 (0.316/0.316)
Observed effects	0.620	0.200	0.167	0.653	2532 ± 1609	0.588 (0.459/0.129)

ASS ± SD: average total sample size and standard deviation
Go to stage 2: amount of cases the trial continues to the second stage (due to Death/CE).

Table 20.7 Power values for different fixed designs

		Power for...			
Scenario	Total sample size	H_1^{Death}	H_1^{CE}	$H_1^{Death} \cap H_1^{CE}$	$H_1^{Death} \cup H_1^{CE}$
Assumed	6600	1.000	1.000	1.000	1.000
effects	2130	0.899	0.899	0.899	0.899
Observed	6600	0.896	0.466	0.464	0.899
effects	2130	0.403	0.163	0.146	0.420

The simulation results clearly demonstrate that the investigated adaptive Bonferroni-Holm design offers a considerable power advantage compared to a fixed design which is based on the wrong planning assumptions. For example, the power to reject at least one of the two endpoint in the observed effect scenario amounts to 0.653 (compare Line 2 of Table 20.6), whereas the corresponding power in the fixed design based on a sample size of 2130 is only 0.420 (compare Line 4 of Table 20.7). However, it also has to be noted that the target power of 0.9 is not met with the adaptive Bonferroni-Holm approach in the observed effects scenarios. This is due to the fact that the deviations from the original planning assumptions are large. As a consequence, the allowed maximal sample size of $5 \cdot 1368 = 6840$ is still too small (remember that the required sample size for the observed effects scenario in the fixed design is already 6600). It is intuitively clear that if the maximal sample size is too low for the underlying parameter constellation, the target power cannot be met.

For the assumed effects scenario, the power to reject at least one of the null hypotheses in the adaptive Bonferroni-Holm approach is given by 0.966 and thus larger than the target power. The average sample size given by 1908, however, is smaller than the required sample size for a corresponding fixed design given by 2130. This clinical trial illustration was originally published by Rauch et al. (2016).

20.5 The LIFE Trial

The LIFE Trial was designed as a randomized, double-blind, controlled clinical trial to compare the effects of losartan and atenolol on cardiovascular morbidity and mortality in patients with diabetes and hypertension (Dahlöf et al. 2002). The primary endpoint was a composite time-to-first-event endpoint combining "myocardial infarction," "stroke," and "cardiovascular mortality." The recruitment time was fixed to 2 years and the minimal follow-up duration to 4 years. A total of 9193 patients were allocated to the two treatment groups in a 1:1 ratio. The LIFE study was planned to detect a decrease in the 5-year event rate of 0.15 in the control arm to 0.1275 in the intervention group with a power of 0.8 at a one-sided significance level of 0.025 (Dahlöf et al. 1997). This effect corresponds to a hazard ratio of 1.19 (control versus intervention) when assuming exponentially distributed event times. The final sample size of 9193 patients was calculated for a group-sequential design with two interim analyses and included a certain recruitment overrun.

The results showed a significant and relevant effect in favor of the intervention. The all-cause hazard ratio for the composite endpoint was given by 1.15 (control versus intervention) which is very close to the planning assumption. The individual components of the composite were also analyzed and reported. The observed hazard ratio for cardiovascular mortality was given by 1.12, the hazard ratio for stroke was 1.33, and the hazard ratio for myocardial infarction was given by 0.93 which corresponds to a small adverse effect. Note that the results related to the components are only of descriptive nature and provide no confirmatory evidence. As the composite endpoint showed a significant effect, a multiple test strategy which uses the composite null hypothesis as a gatekeeper followed by a Bonferroni-Holm procedure to test the components, as proposed in Chap. 17 in Part V of this book, would clearly provide a benefit here. This strategy is investigated in the following.

20.5.1 Gatekeeping Followed by Bonferroni-Holm

As the underlying hazards for the observed effects are not known, we will assume that the observed effects for the composite and the components are given as follows:

$$\theta_{CE} = \frac{0.0325}{0.0273} = 1.16,$$

$$\theta_{Death} = \frac{0.0103}{0.0091} = 1.13,$$

$$\theta_{MI} = \frac{0.009}{0.0092} = 0.98,$$

$$\theta_{Stroke} = \frac{0.0132}{0.009} = 1.47.$$

Again, the corresponding time scale is based on months. The hazard assumptions specified above are chosen such that the cause-specific hazards sum up to the all-cause hazard in each group and such that the hazard ratios roughly correspond with the truly observed effects. For the sake of simplicity, the two planned interim analyses are neglected in the following considerations. The total sample size of 9193 patients provides a power of 0.875 for a composite hazard ratio of $\theta_{CE} = 0.0325/0.0273 = 1.16$ when applied in a fixed design without interim analyses. Under these assumptions, we generated a total of 10,000 random data sets and applied the multiple test strategy using the composite null hypothesis as a gatekeeper followed by Bonferroni-Holm for the components. The power results are provided in Table 20.8.

The simulated power to reject exclusively the composite null hypothesis correctly amounts to 0.875. Moreover, within this multiple test strategy, there is a chance of 0.863 to reject in addition the null hypothesis of one component. This additional confirmatory evidence for at least one component can be generated "for free"

Table 20.8 Power values for different fixed designs

	Power for rejection of the null hypothesis for..			
Total sample size	CE alone	CE and 1 component	CE and 2 components	CE and 3 components
9194	0.875	0.863	0.166	0.003

without the need to increase the sample size. The chances to additionally reject the null hypotheses of two or three components are relatively small with 0.166 and 0.003. However, it should be noted that in the considered scenario, only the component "stroke" shows a considerable effect. The effect for "death" is rather small and based on very small hazards implying a low number of events. The hazard ratio for "myocardial infarction" even points in the opposite direction, and therefore a rejection of the corresponding null hypothesis would correspond to a false decision. In conclusion, the applied multiple test strategy clearly provides a benefit over the standard approach to test exclusively the composite null hypothesis and to provide only descriptive analyses for the components. This clinical trial illustration was similarly published by Schüler et al. (2014).

References

Brenner, M. B., Cooper, M. E., de Zeeuw, D., Grunfeld, J.-P., Keane, W. F., Kurokawa, K., et al. (2000). The losartan renal protection study – Rationale, study design and baseline characteristics of RENAAL (Reduction of endpoints in NIDDM with the angiotensin II antagonist losartan). *Journal of Renin-Angiotensin-Aldosterone System, 1*, 328–335.

Brenner, M. B., Cooper, M. E., de Zeeuw, D., Keane, W. F., Mitch, W. E., Parving, H. H., et al. (2001). Effects of losartan on renal and cardiovascular outcomes in patients with type 2 diabetes and nephropathy. *New England Journal of Medicine, 345*, 861–869.

Dahlöf, B., Devereux, R. B., de Faire, U., Fyhrguist, F., Hedner, T., Ibsen, H., et al. (1997). The losartan intervention for endpoint reduction (LIFE) in hypertension study: Rationale, design, and methods. *American Journal of Hypertension, 10*, 705–713.

Dahlöf, B., Devereux, R. B., Kjeldsen, S. E., Julius, S., Beevers, G., de Faire, U., et al. (2002). Cardiovascular morbidity and mortality in the losartan intervention for endpoint reduction in hypertension study (LIFE): A randomised trial against atenolol. *The Lancet, 359*, 995–1003.

Dargie, H. J. (2000). Design and methodology of the CAPRICORN trial - A randomised double blind placebo controlled study of the impact of carvedilol on morbidity and mortality in patients with left ventricular dysfunction after myocardial infarction. *European Journal of Heart Failure, 2*, 325–332.

The DREAM (Diabetes REduction Assessment with ramipril and rosiglitazone Medication) Trial Investigators (2006). Effect of rosiglitazone on the frequency of diabetes in patients with impaired glucose tolerance or impaired fasting glucose: A randomised controlled trial. *The Lancet, 368*, 1096–1105.

ICON plc. ADDPLAN Software. (2014). Version 6.11. Accessed February 6, 2018. http://www.iconplc.com/technology/addplan/. Accessed April 9, 2018.

Pocock, S. J. (1977). Group sequential methods in the design and analysis of clinical trials. *Biometrika, 64*, 191–199.

Rauch, B., Schiele, R., Schneider, S., Diller, F., Victor, N., Gohlke, H., et al. (2010). OMEGA, a randomized, placebo-controlled trial to test the effect of highly purified omega-3 fatty acids on top of modern guideline-adjusted therapy after myocardial infarction. *Circulation, 122*, 2152–2159.

Rauch, G., Jahn-Eimermacher, A., Brannath, W., & Kieser, M. (2014a). Opportunities and challenges of combined effect measures based on prioritized outcomes. *Statistics in Medicine, 33*, 1104–1120.

Rauch, G., Kieser, M., Ulrich, S., Doherty, P., Rauch, B., Schneider, S., et al. (2014b). Competing time-to-event endpoints in cardiology trials: A simulation study to illustrate the importance of an adequate statistical analysis. *European Journal of Preventive Cardiology, 21*, 74–80.

Rauch, G., Wirths, M., Kieser, M. (2014c). Consistency-adjusted alpha allocation methods for a time-to-event analysis of composite endpoints. *Computational Statistics and Data Analysis, 75*, 151–161.

Rauch, G., Kunzmann, K., Kieser, M., Wegscheider, K., & Eulenburg, C. (2018). A weighted combined effect measure for the analysis of a composite time-to-first-event endpoint with components of different clinical relevance. *Statistics in Medicine, 37*, 749–767.

Rauch, G., Schüler, S., Wirths, M., Englert, S., & Kieser, M. (2016). Adaptive designs for two candidate primary time-to-event endpoints. *Statistics in Biopharmaceutical Research, 8*, 207–216.

Schüler, S., Mucha, A., Doherty, P., Kieser, M., & Rauch, G. (2014). Easily applicable multiple testing procedures to improve the interpretation of clinical trials with composite endpoints. *International Journal of Cardiology, 175*, 126–132.

The CAPRICORN Investigators (2001). Effect of Carvedilol on outcome after myocardial infarction in patients with left-ventricular dysfunction: The CAPRICORN randomised trial. *The Lancet, 357*, 1385–1390.

Index

G. Rauch et al., *Planning and Analyzing Clinical Trials with Composite Endpoints*, Springer Series in Pharmaceutical Statistics,
https://doi.org/10.1007/978-3-319-73770-6

W

MIX
Papier aus verantwortungsvollen Quellen
Paper from responsible sources
FSC® C105338

If you have any concerns about our products,
you can contact us on
ProductSafety@springernature.com

In case Publisher is established outside the EU,
the EU authorized representative is:
Springer Nature Customer Service Center GmbH
Europaplatz 3, 69115 Heidelberg, Germany

Printed by Libri Plureos GmbH
in Hamburg, Germany